Springer Polar Sciences

This Series is now being indexed by SCOPUS.

Springer Polar Sciences is an interdisciplinary book series that is dedicated to research in the Arctic, sub-Arctic regions, and the Antarctic. In recent years, the polar regions have received increased scientific and public interest. Both the Arctic and Antarctic have been recognized as key regions in the regulation of the global climate, and polar ecosystems have been identified to be particularly susceptible to the ongoing environmental changes. These changes are having widespread implications for human communities, businesses, and governance systems and are interacting with demographic shifts, globalisation, resource development, cultural change, territorial disputes, and growing calls for self-determination in some regions. Consequently, the international efforts in polar research have been enhanced considerably, and a wealth of new findings is being produced at a growing rate by the international community of polar researchers and those who live in the region.

Springer Polar Sciences aims to present a broad platform that will include state-of-the-art research, bringing together both science, humanities, and perspectives rooted in indigenous and local knowledge to facilitate an exchange of knowledge between the various polar science communities. The series offers an outlet to publish contributions, monographs, edited works, conference proceedings, etc. Topics and perspectives will be broad and will include, but not be limited to, climate change impacts, climate change policy, environmental change, polar ecology, governance, health, economics, indigenous populations, tourism, resource extraction activities, and research design in polar regions. Books published in the series will appeal to scientists, students, polar researchers, community leaders, and policy makers.

Douglas C. Nord
Editor

Contemporary Nordic Public Policies

Degrees of Exceptionalism and Inner Diversity

Springer

Editor
Douglas C. Nord
Department of Political Science
Umeå University
Umeå, Sweden

ISSN 2510-0475 ISSN 2510-0483 (electronic)
Springer Polar Sciences
ISBN 978-3-032-11291-0 ISBN 978-3-032-11292-7 (eBook)
https://doi.org/10.1007/978-3-032-11292-7

This Springer imprint is published by the registered company Springer Nature Switzerland AG
The registered company address is: Gewerbestrasse 11, 6330 Cham, Switzerland

To that which we share, and
to that which makes each distinctive

Preface

This volume has its origins and is rooted in an ongoing paradoxical situation. Why, on the one hand, should the nations of Scandinavia—or for that matter, the broader Nordic region—continue to be regarded as some of the most prosperous, democratic, and socially cohesive countries on the globe of and yet, on the other hand, remain among some of the least observed, reported, and scrutinized societies of the modern world. Various global rankings originating from the United Nations, reputable think tanks, and a number of major research universities regularly point to the fact that the Nordic nations enjoy some of the highest standards of living, offer some of the best educational and training services to their citizens, and operate as major centers for innovation and competitiveness. Others suggest that these same societies top any list with regard to environmental awareness, gender equality, the observance of human rights, and freedom of expression. They are also seen as having some of the best architects, mystery writers, and restaurants on the globe. As a result, many analysts have suggested that these countries are among the "best places in the world to live" and have some of the "happiest" and most satisfied societies on earth.

Despite these glowing assessments, the countries of the Nordic region continue today to be largely uninvestigated by the media, academic researchers, and the general public. Few news stories carried by the world press originate from the area. Perhaps only a couple of times a year will these countries feature on any of the front pages of leading newspapers outside the region. Despite having some of the most breathtaking scenery in the world, you will not find many travel guides to the region. While there may be a wealth of offerings on France, Italy, the United States, China, and Brazil, one will struggle to find more than a just a handful that attempt to explore, in detail, any of the Nordic countries. Nor do the research and educational communities outside of the region endeavor to investigate key features of these nations. A review of major global research indexes finds only a limited number of citations focusing on Scandinavian/ Nordic topics by authors outside the region. Similarly, one discovers only scant references to the area in the curriculum of schools and universities across the globe. Despite touting the virtues of the Nordic community, little is done to actually explore the history, contemporary culture or public policies of these nations. Reflecting on this reality, British journalist Michael Booth has sardonically noted in his book

The Nearly Perfect People that: "The truth is that we learn more from our schoolteachers, televisions and newspapers about the lives of remote Amazonian tribes than we do about actual Scandinavians and how they actually live."[1] For many across the globe, Scandinavia remains *terra incognita.*

Reasons Why the Nordics Remain Unknown

It seems to me that there are at least five factors that help explain why the Nordics continue to remain unknown to much of the rest of the world. The first of these is the area's small population.

Although the Nordic region is quite large in territory—the seventh largest on the globe—its residents are few in number. The 28 million who inhabit it represent less than 1% of the world's total population. As a consequence, the Nordics can be easily overlooked at times. Like many other large but sparsely populated states like Canada and Australia, they have to compete with other more highly populated countries for attention and focus. With the continuing decline in coverage by the global media and press of "small societies," they tend to be routinely pushed to the back of world's consciousness. The assumed homogeneity among the Nordics also does not help garner attention. The peoples and societies of Norway, Sweden, Denmark, Finland, and Iceland are frequently lumped together and seen to be interchangeable in the eyes of many tourists, business leaders, and government officials who do not bother to turn their eyes toward the North.

A second and related factor which underlies this lack of familiarity is the Nordic region's relatively remote location. Much of Scandinavia lies "above the fold in the map" of Europe, and when Iceland and Finland are added in, an even greater sense of distance from the center emerges. Even with modern advances in communications, technology, and transport, the Nordic region seems for many to be on the periphery of the "known world" with few other neighbors and a limited number of regular lines of contact with societies to the south. Added to this sense of relative isolation is the region's challenging climate. The cold and darkness of the Nordic area in the winter is legendary and frequently seen as a barrier to both commercial and societal interaction with others. Added to this is the frequently cited reserved character of the residents which also tends to undercut interest and communication from the outside.

A third contributing factor to the lack of attention accorded to the Nordics is their relative low profile within the history and culture of the broader global community. Their societal experiences and contributions are frequently overlooked. The two exceptions seem to be the impact of the Vikings on the development of Europe and the arrival of Scandinavian immigrants to the New World. Both of these events make it into the history books and popular culture of much of the rest of the globe. However, unfortunately, they tend to overwhelm and obscure many of their more

---

[1] Michael Booth, *The Almost Perfect People: Behind the Myth of the Scandinavian Utopia.* New York: St. Martin's Press, 2014 p. 8.

recent contributions. As a consequence, Nordic residents are frequently seen by those from the outside as "people of the past" confined to memories and experiences of a distant historical era and not noteworthy with regard to contemporary concern.

A fourth factor that inhibits a greater awareness of the Nordic community has been the tendency to compress what is "known" of its members into stereotypical images. For a good deal of the world, all Scandinavians are tall, blond, blue-eyed types that commune with nature and are beholding to an extensive welfare state. They enjoy a relatively high standard of living, pay high taxes to their governments, and experience rather low-keyed and inward-looking lives. They supply serviceable furniture and consumer goods as well needed raw materials and high-quality industrial products. The Nordics are deemed to be relatively silent people who avoid conflict with others and pursue foreign and defense polices focused on neutrality or non-alignment. Sometimes to this collective imagery is added the somewhat judgmental observations that they have problems with their drink and that they are often shy and dull in comparison with other societies. While each of these stereotypes may have elements of truth within them, they combine to produce an external assessment that suggests that the Nordics are either "too good to be true" or not challenging enough for any extended consideration. Frequently it is heard that they just do not face the same problems encountered by the rest of the world.

This leads to the fifth factor that limits focus and attention from the outside world. Ultimately, much of Scandinavia is seen as an admirable but inherently uninteresting region. It must compete for the world's regard with much more varied and colorful communities. Like Canada which has been referred to as the "good, gray country," Scandinavia must contend with a public image of being clean, nice, but not terribly newsworthy. It tends to lack the color and controversy which often grabs the attention of the outside observer who is daily bombarded with a variety of competing images and storylines. As a consequence, the region remains for many—up there, out of sight, out of mind. It often occupies a position of a being a nice afterthought once more pressing and troublesome images from the rest of the world have been considered.

Reasons Why the Nordics Should Be Known Better

While the above factors combine to create a formidable barrier to an enhanced awareness and a better understanding of the Nordic area, it does not mean that no effort should be made to overcome them. In fact, one of the goals of this book is to assist the process of giving greater priority to the region. Careful consideration suggests that there are a number of reasons why the Nordics should be accorded increased prominence and attention by the global community. The rationale for doing so builds on some of the features of the societies outlined above but offers an alternative lens or perspective from which to view them.

It is true that the Nordic nations are not large states with regard to population. However, these "small states" display a somewhat surprising degree of influence within the global community. Over recent decades they have provided leadership and inspiration for a variety of international organizations and their agendas including those of the United Nations, the European Union, and NATO. Their capital cities have frequently hosted major international conferences and negotiations focusing on the environment, human rights, and international trade and development.

In the business and commercial spheres, the Nordic nations have not only produced major global commercial players like Volvo, Moller-Maersk, H&M and Carlsberg but they have also operated as important centers of research and development. In similar fashion, they provide a variety of cultural products that have their impact around the world. Lego, Scandinavian mystery novels, Moomin characters and songs from ABBA are eagerly sought after by consumers across the globe. Like other "small states" including Australia, New Zealand, Switzerland, and Singapore the Nordics exert broad influence and can have a major impact despite their relatively small populations. This ability to "punch above one's weight" is increasingly recognized as a feature of global power in the international arena and something that need to be attuned to and taken into consideration by other larger players.

Nor is the supposed remoteness and peripheral nature of the Nordic communities entirely to be accepted. Contemporary events point to the fact that they are not as distant or marginal as some have suggested in the past. The war in Ukraine, the expansion of NATO, and the emergence of the Arctic as a zone of potential global cooperation or conflict all suggest that the Nordics are becoming more central to international concern and activity. Their roles within the European Union have also become increasingly significant and consequential over recent decades. So too has the Nordics community's participation in global trade and markets. Also, despite the challenges of a demanding climate and the recent impact of the Covid pandemic, global tourism flows to the region have been steadily rising as an increasing number of European, Asian, and North American travelers see the Nordic area as an exciting destination. The region should no longer deemed to be on the "margins of things" but, instead, be seen as moving steadily towards its center.

A similar reorientation is now taking place in the minds of many regarding the Nordics' role in shaping the future. Increasingly, they are no longer being viewed as "peoples of the past," but rather as "leaders of the future." This can be seen in the roles they have played both within the commercial and technological arenas as well as in the broader domain of modern living. Nordic companies like Nokia and Ericsson were among the first to pioneer cellphones, and they and other firms from the region continue to play leading roles in developing new products and technologies related to communications, transport, energy production, and artificial intelligence. Today, the Scandinavians are generally recognized as being in the global forefront in developing both environmentally friendly and sustainable goods and services. Firms ranging from IKEA to Bang and Olufsen are praised for their simplicity, functionality, and quality of their design. From a somewhat different perspective, the growth and development of the Nordic welfare states have also received similar praise. Their abilities to provide quality education and training, available child care, affordable public transport, and broad support for the aged and infirmed of their communities are increasingly recognized by other nations as models to be followed and emulated.

While the Nordics continue to receive positive notices regarding these efforts, they are also increasingly being portrayed as communities encountering a variety of societal problems and challenges not unlike those found elsewhere in the world. These include matters regarding the integration of new immigrants and refugees into their societies, the increasing threats of climate change and environmental hazards, growing levels of crime and violence in their urban centers, and a variety of social and economic pressures faced by both its youth and aged populations. Similarly, within the Nordic societies, there is a perceived need for greater attention to be focused on confronting the economic challenges of supporting a mature social welfare state as well as addressing growing defense and security needs. With these matters on their agenda, the Nordic nations are increasingly seen as not being that dissimilar from many other post-modern societies found in Europe, North America, and East Asia. As such, the supposed halo's glow of being "the almost perfect" societies has faded somewhat and the Nordics are now subject to more critical examination and comparison with others. Today, the degree of their uniqueness or exceptionalism is more open to question than ever before and the diversity among them in their formulation of public policy responses more evident.

Finally, it is becoming increasingly clear that the Scandinavian states should no longer be easily consigned to the convenient category of the "good, gray region" to the North. The Nordics in the first quarter of the twenty-first century have gained new prominence and regard throughout the world for a number of reasons some of which have been outlined above. Externally, they are now perceived as playing more a more significant role in global economic, political and cultural affairs than has been the case in the past. Internally, they can be seen as being more colorful, diverse and innovative than previously assumed. In 2025, the Nordic nations can now be viewed as facing some of the same challenges encountered by other contemporary societies. The manner of their approach to these, and the policy measures that they pursue in addressing them, should be of interest and consequence for a wider global audience.

Purpose and Structure of the Book

The primary purpose of this book is to provide a contemporary introduction to the Nordic community and to help familiarize its readers with some of the main features and concerns of the area. In so doing, specific attention is given to examining the political culture of the region and some of the major public policy efforts undertaken by its members in recent years. An effort is made to consider both the important similarities existing between the constituent nations as well as certain significant differences in thinking and approach that are also evident. With this in mind, an undertaking is made both to describe and analyze some of the important common features of the Nordic Model that seem to be operative within each of these societies as well as to account for what may appear to be clear discontinuities in its application and implementation among them. Overall, the volume seeks to offer a clear and focused consideration of the Nordic community today and some of its significant accomplishments and challenges.

While this book will be certainly of interest and use to residents of the Nordic states, it is not intended that its readership should be limited to that area alone. Instead, one of its chief audiences should be those from the broader global community who may know little about the Nordic region but who desire to learn more. As noted above, there have been limited opportunities to acquaint those outside the area with some of its most salient features. Thus, this volume endeavors to contribute to both expanding international knowledge and awareness of the Nordic region as well as, perhaps, correcting certain misunderstandings and stereotypes that continue to exist. Moreover, by focusing on the political culture and public policy choices of the Nordic area, the book seeks to provide opportunities to draw direct comparisons between the political systems found there and those experienced in the reader's own country. It also allows for a consideration of how different societies may choose to respond to common concerns and problems and the extent to which Nordic "exceptionalism" exists.

The book is composed of a series of individual essays linked together by a common interest in the Nordic community of today. The initial component of the volume incorporates two chapters that offer a framework for analysis. The first chapter begins with a short discussion of the constituent members of the Nordic community and some of their chief characteristics and common history. It seeks to offer a broad overview of their social, political and economic development and some of their efforts to develop regional cooperation in the post-World War II era. It also introduces the idea of a Nordic Model and the ongoing questions of both Nordic "exceptionalism" and "inner diversity" reflected in their public policies. This beginning chapter is followed up with a second that examines the evolution of public policies within the region over the past several decades. It identifies three defining eras or phases of their development from the 1960s to the present time and suggests how each has directed and shaped public and governmental thinking about the how such policies should be fashioned. This latter chapter also examines some of the more pressing issues and problems that the region faces today and suggests that an adequate response to them may require drawing upon both long-established and new approaches to public policy making.

The second component of the book highlights several thematic concerns of contemporary public policy within the Nordic community. They include matters related to the evolution of political culture and the welfare state, the impacts of liberalization and globalization on the region, the nature of changing security and foreign policy requirements, and the significant challenges posed by climate change and associated environmental challenges. Each of these thematic areas are explored, in turn, by individual authors—or a team of authors—drawn from throughout the Nordic region who are well-acquainted with the particular subject matter under consideration. In each of their essays, they endeavor to examine the current and likely future public policy paths that are being pursued by the members of the Nordic community with regard to their identified policy area. Each of the chapter authors also seek to place their discussion of specific public policy choices within the context of considering both the degree of Nordic exceptionalism and inner diversity found in each instance.

The third and final component of the book provides a summary of the previous discussions and a statement of conclusions that can be reasonably drawn from them. It points to certain cross-cutting themes that have emerged among the several substantive chapters. It identifies the type and extent of both internal and external influences that seem to be operative in the various cases detailed. It endeavors to outline the degree of commonality and diversity and that can be found in such contemporary Nordic public policy formulation. It also seeks to draw comparisons with the efforts of other communities in the broader international setting and to suggest why the Nordic experience may be one that the rest of the global community may want to give closer examination and consideration to in the future.

As indicated earlier, this volume benefits from the perspectives and insights of scholars and researchers from across the Nordic region representing several different academic disciplines. It is believed that by providing such a broad comparative and interdisciplinary orientation, the reader can gain a better understanding of both the common influences that contribute to Nordic public policies throughout the area as well as some of the distinctive features seen within specific national communities. These ideas of exceptionalism and inner-diversity are found at the center of the discussions offered. They help us to better understand both the common threads that continue to bind together these societies in today's world as well as the some of the differentiating forces that contribute to a diversity of thinking and action found within the region that are too often overlooked. It is hoped that by providing such perspectives, an enhanced understanding of the Nordic community and its public policy choices can be advanced.

Umeå, Sweden Douglas C. Nord

Acknowledgments

This volume would not have been completed without the helpful assistance of several individuals. First, let me acknowledge the critical roles which three of my Umeå colleagues played in its origins, development, and completion. I wish to thank Gunnel Gustafsson for instigating the plan of bringing together scholars from across the Nordic region to consider what common and dissimilar features of public policy were evident in their separate societies. Gunnel was the originator of the idea for the subtitle for of the book—degrees of exceptionalism and inner diversity—and it was she that kept us focused on this challenging but important perspective throughout. My thanks are directed as well to Anders Lidström who helped develop the structure of the volume, noting early the need to provide both an introductory overview of the Nordic region and a specific discussion of the phased evolution of public policy development within it. Anders was also of tremendous help in identifying key scholars who might contribute to the volume and in organizing the original symposium from which the book sprang. A debt of gratitude is also owed to Niklas Eklund who played a critical role in ensuring that the creative effort continued over the subsequent months. Niklas was able to intervene when some of our original participants in this research endeavor were forced to drop out of the project and helped identify excellent replacements. He also ably assisted me in keeping us all focused on our common goal and kept us on deadline. All three of my colleagues have not only contributed insightful chapters to the volume, but have provided me valued counsel and advice along the path toward its publication.

Second, let me also thank each of the other individual authors who contributed chapters to the volume. It would not have emerged without their expertise and cooperation. The book reflects not only the variety of their backgrounds but also their willingness to accept a common challenge. Collectively they represent nearly a dozen different academic disciplines and each of the societies found within the Nordic region. This is unusual for an edited volume of this sort which often tends to be the product of a single disciplinary perspective provided by authors usually only from a subset of the region. This book was intended from the outset to be broad and embracing assisting both experts and newcomers to the region to better understand the variety and complexity of Nordic public policymaking as well as common themes and continuing connections. I thank each of the authors for their contributions to the effort. I also wish to acknowledge their willingness to work with me in bringing forth and highlighting these important dimensions. They responded with patience and good humor to my repeated calls to stay focused and on schedule and did not grumble too much when I presented them with suggested revisions and edits. I appreciated your thoughts and feedback throughout the editing process. It has reinforced my belief in the value of interdisciplinary inquiry and collaboration.

Third, a word of appreciation needs also to be directed towards Lyndon Olson, the former U.S. ambassador to Sweden and long-time friend and supporter of Umeå University. In some ways, his continuing interest in Sweden and the broader Nordic region was an impetus for the whole effort. Some two and half decades ago he was scheduled to deliver a speech at the university giving his impressions of the similarities and differences between public policy development in Sweden and his own country. Unfortunately, the 9/11 crisis intervened and he was unable to offer the presentation. However, he was kind enough to forward his remarks and they served as a lingering reminder to all who framed this inquiry of the importance of not only comparing the public policy efforts of the Nordic states with one another, but also considering how they relate to other societies in Europe, North America, and Asia. When it came time to hold the symposium that gave rise to this book, Ambassador Olson was gracious enough to accept an invitation to attend and participate in it. He not only patiently listened to the academic papers that were presented there but also offered some sage comments and advice for the authors to consider. He helped us all to remember that public policy has to be implemented by government departments and those within them, and that these various actors can also have an impact on their evolution.

Finally, an acknowledgment should be given to the generous support provided by a variety of entities for the development and compiling of this volume. A major source of funding for the effort has come the Scandinavian Policy and Political Studies Foundation at Umeå University. It covered the most of the expenses associated with holding the original symposium and the subsequent editing of this volume. Additional financial assistance came from the Department of Political Science and the Faculty of Social Sciences at Umeå University. The Arctic Centre at the university was also an important contributor to the effort. To all of these supporters, a hearty thanks.

Umeå, Sweden Douglas C. Nord

Contents

Editor and Contributors

About the Editor

Douglas C. Nord is Visiting Professor of Political Science at Umeå University, Sweden. He is an established scholar in the fields of comparative politics and international relations. He is interested in the processes of policy learning among countries with a particular focus on immigration, the environment and education. He has also written extensively of the relations between the countries of the circumpolar North and on the emergence of the Arctic as a central concern of contemporary international politics. Nord has studied the development of the Arctic Council over the past three decades. He is the author of *The Arctic Council: Governance within the Far North; The Changing Arctic: Creating a Framework for Consensus Building with the Arctic Council* and the edited volume *Leadership for the North: The Influence and Impact of Arctic Council Chairs.* His most recent book is *Sister City Diplomacy* which examines local community engagement in U.S.-Russian relations from the Cold War to the present era. Nord has been a Fulbright Senior Research Scholar and the Chair of the Scientific Advisory Board for NordForsk's Centers of Excellence in Arctic Research. He edited the volume, *Nordic Perspectives on the Responsible Development of the Arctic*, which examines contemporary research efforts in the region.

Contributors

Monika Gabriela Bartoszewicz The Arctic University of Norway, Tromsø, Norway

Tina Øllgaard Bentzen Roskilde University, Roskilde, Denmark

Lennart Joe Brunkert Leuphana University, Lüneberg, Germany

Haldor Byrkjeflot University of Oslo, Oslo, Norway

Romain Chuffart University of Akureyri, Akureyri, Iceland

Niklas Eklund Umeå University, Umeå, Sweden

Anastasia Emelyanova University of Oulu, Oulu, Finland

Birgitta Evengård Umeå University, Umeå, Sweden

Gunnel Gustafsson Umeå University, Umeå, Sweden

Keith Larson Umeå University, Umeå, Sweden

Anders Lidström Umeå University, Umeå, Sweden

Magnus Lindmark Umeå University, Umeå, Sweden

Douglas C. Nord Umeå University, Umeå, Sweden

Bi Puranen Institute for Future studies, Stockholm, Sweden

Håkan T. Sandersen Nord University, Bodø, Norway

Paul Schmidt Umeå University, Umeå, Sweden

Olof Stjernström Nord University, Bodø, Norway

Johan Strang University of Helsinki, Helsinki, Finland

Marte Winsvold Institute for Social Research, Oslo, Norway

Corine Wood-Donnelly Nord University, Bodø, Norway

Part I
Overview

Chapter 1
An Introduction to the Nordic Region

Niklas Eklund and Gunnel Gustafsson

Abstract This volume seeks to examine the nature of contemporary Nordic public policies from the vantage point of how they are "exceptional" as well as demonstrating features of their "inner diversity." We use these terms to capture their dual nature. On the one hand, they present a distinct character that sets them apart from other public policy formulations found elsewhere in the world. They reflect a unique Nordic orientation in their creation and in the way they are implemented. On the other hand, they are not exactly uniform in character. Despite this shared Nordic heritage, one can discern specific differences in how each country of the region advances its own policy preferences and agendas. The essays contained within this book collectively seek to address both halves of the question before us. How do Nordic public policies of the contemporary era reflect both shared traditions and preferences as well as distinctive national differences and priorities? Together both elements help to provide us with an enhanced understanding of the public policy choices made within the Nordic region today.

Keywords Geography · Population · History · Institutions · Cooperation

The purpose of the present chapter is to lay the foundation for a nuanced understanding of the Nordic region. It can be seen as an update for those who are already familiar with the countries involved or, for others, as a first introduction to the region and a means to become more informed regarding different degrees of exceptionalism and inner diversity within the public policies of this part of the world. Basic information is offered regarding the geographic location and history of the Nordic region, its economic foundations as well as other data about the five countries and three autonomous areas involved. The chapter provides details concerning population size, degrees of urbanization, languages and so on. It also outlines certain similarities and differences in governance structures within the region and describes some of the joint Nordic actions that have been undertaken to alleviate regional problems,

N. Eklund (✉) · G. Gustafsson
Umeå University, Umeå, Sweden
e-mail: niklas.eklund@umu.se

D. C. Nord (ed.), *Contemporary Nordic Public Policies*, Springer Polar Sciences,
https://doi.org/10.1007/978-3-032-11292-7_1

increase economic cooperation and meet shared challenges and opportunities over several decades. The chapter concludes with a discussion of the concept of the Nordic Model and its various interpretations.

1.1 Characteristics of the Nordic Region

The five countries, Denmark, Finland, Iceland, Norway, and Sweden, as well as the three autonomous areas of Greenland, the Faroe Islands, and Åland collectively constitute the Nordic area. Greenland and the Faroe Islands belong to Denmark and Åland, an island situated in the Baltic Sea between Finland and Sweden, belongs to Finland. Nowadays, the terms "Nordic" and "Scandinavian" are perceived by many as synonyms, but Nordic is the term that will be used when referring to the relevant region shown on the map (Fig. 1.1).

Table 1.1, outlines some of the main characteristics of the separate components of the region. As can be seen, the countries and autonomous regions vary considerably with regard to size and population. Greenland is the largest island in the world spanning as much as 2,166,086 km^2 whereas the Faroe Islands and Åland, in comparison, are very small covering only 1581 km^2 and 1396 km^2 respectively. Sweden with its 447,435 km^2 has the largest area among the five countries and Denmark the smallest

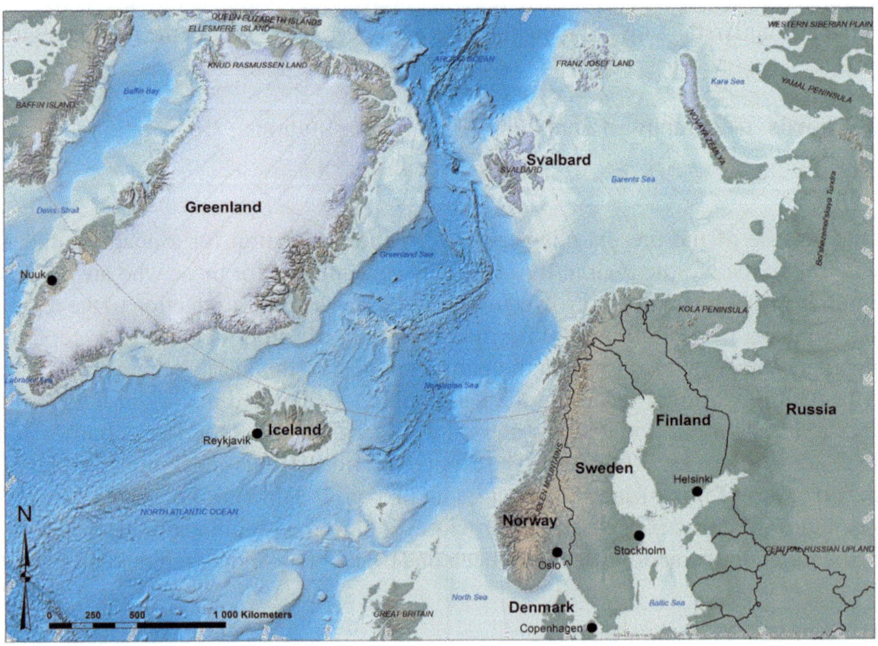

Fig. 1.1 The Nordic region

Table 1.1 Characteristics of the countries and autonomous areas in the Nordic region

Countries and autonomous areas	Area in km^2	Population	Capital city	Currency	Population in cities (%)	Immigrants (%)
Denmark	43,561	5,840,045	Copenhagen	Danish crown	33.13	10.5
Finland	338,430	5,533,793	Helsinki	Euro	40.11	6.1
Iceland	103,492	36,892	Reykjavik	Icelandic crown	62.11	14.1
Norway	323,781	5,367,580	Oslo	Norwegian crown	31.20	14.4
Sweden	447,435	10,379,295	Stockholm	Swedish crown	35.77	18.1
Greenland	2,166,086	56,000	Nuuk	Danish crown		
Faroe Islands	1396	52,934	Tórshavn	Faroese crown		
Åland	1581	30,129	Mariehamn	Euro		

Source https://www.norden.org/sv/information/fakta-om-de-nordiska-landerna

with its 43,561 km^2. Also, the number of inhabitants vary considerably between the constituent parts of the Nordic region. The countries of Denmark, Finland and Norway have similarly sized populations, while Sweden's in nearly twice as large. In comparison, Greenland, the Faroe Islands and Åland are quite small.

Most of the languages in the Nordic region belong to the Indo-European family.[1] This is the case for Norwegian, Danish, Icelandic, and Swedish and Faroese. The languages of the three Scandinavian countries (Norway, Denmark and Sweden) are relatively similar and can be mutually understood. Pronunciation does, however, sometimes make it difficult for Swedes and Norwegians to understand Danish and vice versa. There are also considerable dialectal differences within each Scandinavian language. Faroese is related to Icelandic, and these two languages are also often hard for the other Nordics to understand. In Finland, there are two official languages, Finnish and Finnish-Swedish. The latter is closely related to Swedish, and more of a dialect than a different language. Finnish, however, belongs to the Finno-Ugric language family and cannot be easily understood by the rest of the Nordics. This is also the case with Greenlandic, which is related to the Inuit languages spoken in Canada and Alaska.

In addition, there are minority languages and the most significant of these are the Sami languages spoken in Norway, Finland, and Sweden. A current estimation of the numbers of Sami people, historically sometimes referred to as "Laplanders," is

[1] For further information about languages and indigenous groups, see for example: Nordic Council of Ministers (2025); Truth Commission for the Sami People (2025); Nordnorge.com (2025); Minority Rights Group (2025); Nordiskt samarbete (2025)

that there are 20–40,000 in Sweden, 50–65,000 in Norway, and 8,000 in Finland. As many as 40–45% of them speak both a Sami dialect and their national language. In each country, there is a Sami parliament, elected by the Sami to safeguard their interests but with some variation in political and administrative mandates. In Finland, the first Sami parliament was established in 1972, in Norway in 1989, and in Sweden in 1993. Likewise, a variety of other languages are used among the increasingly multicultural communities within the Nordic countries today, especially in the major urban areas of the region. These language use patterns provide both examples of a shared inheritance as well as diversity with the region. They also capture the growing immigrant presence found within the Nordic societies.

When representatives from different Nordic countries meet formally to discuss matters of mutual concern it is still considered important to use one of the three languages defined as the "working languages" of the region. These are Danish, Norwegian, and Swedish. As some of the finer points of discourse may not always be understood by all participants, interpretation services are often offered. Sometimes English is also used for presentations and discussions of issues. The increased use of English in such meetings is also reflective of how that language has become a "working language" in the European Union (EU). However, Finnish, Danish and Swedish are still among the twenty-four official languages of the EU.

The Nordic region also reflects some diversity in its economic foundations. While all the countries have developed from primarily natural resource-based economies, they vary today in the extent to which they have retained this component. Some like Norway, Finland and Iceland are still very dependent on these types of commodities while Sweden and Denmark have become considerably more diversified in their economic base. Some of the countries (Denmark, Sweden and Finland) have become full members of the European Union, while others (Norway and Iceland) have remained outside. Yet even among those who have joined, there is diversity in their choice of currency. As Table 1.1 points out, Finland has adopted the Euro whereas Sweden and Denmark have retained their crowns. This provides another example of the inner diversity to be found within the region.

For many foreign observers, the Nordics are perceived as distinctive or "exceptional" regarding their welfare states. They provide their inhabitants with generous benefits such as free health care and free education for children as well as tax-funded social insurance systems and additional financial assistance for those who need such support (Further details are provided in Chaps. 2 and 4 of this volume). Another frequently observed characteristic of the Nordic societies is their relatively large public sectors—this despite a strong commitment to private ownership and private enterprise. High levels of taxation are also to be noted (For further information see Chap. 5 of this book). Overall, the Nordics can be seen to pursue similar economic and social policies at home and shared foreign policy objectives abroad. These commonalities suggest that they are like-minded in their thinking and approach to public policy, yet, as will be seen in subsequent chapters of this book, there are also significant degrees of distinction and inner diversity among them.

In addition, the Nordics have similar systems of democratic governance. Their members of parliament are elected in proportion to their overall support within the electorate and operate within multi-party-political systems. The number of these political parties has increased over time. In Denmark, Norway Finland, and Sweden the relatively "new" political parties reflect a world-wide rise of support for right-wing populism. However, in Iceland the new Icelandic Party is characterized by left-wing populism (Ljunggren, 2023). The increased number of Nordic political parties often makes it a time-consuming process to form governments after an election. Despite these new political trends, the Nordic countries continue to rank high on indexes that measure both levels of democracy and freedom of expression. With regard to these dimensions of society, the Nordics countries continue to stand apart as exceptional in their commitment to democracy (See Chap. 3 for additional details).

There are, however, some differences among these countries regarding their specific forms of governance. Denmark, Norway, and Sweden are monarchies, although the powers of their royal families are minimal and mainly ceremonial in nature. Finland and Iceland, on the other hand are republics. The national parliaments in each of the Scandinavian countries are the law-making bodies with executive powers being vested in their prime ministers and their respective government offices. In the specific case of Finland, the parliament is the law-making body, while the executive power is shared between the president and the prime minister. In Iceland, the parliament named the "Allting" is the lawmaking body and the president has mainly ceremonial duties. The "Allting" was established already in the year of 930 and is the oldest still functioning parliament in the world (Lugmayer, 2009).

Another similarity between the Nordic countries is the importance of subnational units with decision making power, i.e., municipalities and regions. They have the responsibility for delivering, among other things, health care, education, housing, childcare and social services to senior citizens. These levels of government "account for around two-thirds of all public expenditure and employ around two-thirds of all public employees." As such, "these local authorities are agents of the welfare state as well as local decision-makers in their own right" (Baldersheim & Ståhlberg, 1994). This is an important commonality and an "exceptional" feature of public policy formulation and implementation within the Nordic community. Yet the specific size, organizational structure and mandate of these local actors can vary significantly among the five countries revealing a certain inner diversity within the region.[2]

During the first two decades of the twenty-first century there have been a flurry of efforts at public sector reform in the Nordic countries. Denmark, Sweden, and Finland have gone through various regional reforms, mainly driven by the need to provide more efficient health care and services delivery (See Chaps. 5 and 6 of this volume). Likewise, significant changes have been made at the level of local government in Norway, Finland, Greenland, Iceland and in the Faroes. Whereas the success of these innovations has varied, the drivers of such reform efforts have been similar. Changes in mobility patterns and labor markets, ageing populations, increased urbanization and the emergence of different needs and expectations among a new generation of citizens within the Nordic countries are all having an impact. In addition,

[2] The number of municipalities among the Nordic countries is: Denmark 98, Sweden 290, Norway 356, Finland 309, Iceland 98, Fareo Islands 30, Greenland 5 and Åland 16.

renewed pressure on the local and regional levels of governance have coincided with the introduction of certain public policies inspired by the New Public Management paradigm (Hood, 1991; Pollitt & Bouckaert, 2000). These have increased the role of economic efficiency as the marker of overall public sector efficiency. Likewise, in the 2020s growing concerns about the adequacy of public health, education and the need for a renewal of military and civil security structures have also reemerged as drivers of public policy reform across the Nordic region (Chapter 7 of this book provides examples of this).

1.2 A Brief Overview of the History of the Nordic Region

The common history of the countries of the Nordic region dates far back in time with Leif Eriksson (970–1022) serving as an emblematic figure of this shared tradition. He was born in Iceland, grew up in Greenland and spent part of his youth in Norway. He was also an adventurer who reached American shores long before Christopher Columbus. During his time, the western portion of the Nordic area became a kingdom called the "Norwegian Realm." The kings of Norway at that time controlled the Faroe Islands, Iceland and Greenland and built close ties to the Viking kings of Denmark and Sweden. Despite the tumultuous spirit of the Vikings, they managed to foster a community among themselves. At the same time, they demonstrated aggression and colonial tendencies toward other parts of Europe.

The ensuing Middle Ages witnessed political competition and military strife not only between the burgeoning kingdoms of the North but also with the Hanseatic League and other communities in what is today Germany, Poland, and the Baltic states. A unifying Nordic mindset seems to have returned, if temporarily, with the establishment of the Kalmar Union in 1397. The intention behind its formation was to safeguard and strengthen the Nordic region during a time when it, in many respects, began to lag other parts of Europe, militarily as well as scientifically. The Union, under the auspices of the Danish crown, encompassed Sweden (at that time including Finland and Åland) and Denmark (then including Norway, Iceland, Greenland, and the Faroe Islands). The Orkney and Shetland islands were also for some time part of the Kalmar Union as they belonged to Norway when the union was formed and because Norway had gained these islands during Viking times.

Many observers from the outside have often assumed that the Nordic region has always been a peaceful part of Europe. Historically, however, the region has fluctuated between internal cooperation and conflict. The Kalmar Union was riddled with political and religious tensions and challenges. Conflict between the major players, Denmark-Norway, and Sweden-Finland increased over time. In a final attempt to assert his right to dominate the Kalmar Union, the Danish king invaded Sweden-Finland with his military forces reaching as far north as Stockholm. What ensued was something that historians have referred to as the Stockholm Blood Bath, in which the Danish king in 1520 ordered public beheading of a group of Swedish nobles. Instead of reaching his goal of solidifying power within the region, the Danish king's actions precipitated increased opposition to Danish influence in Sweden. A major contender for the Swedish crown, Gustav Vasa, led a growing military resistance in that country and drove the Danish army back south. When Gustav Vasa took the Swedish throne in 1521, Sweden-Finland simultaneously became a major Protestant power in Europe, something which would throw long political shadows over vast parts of the continent for centuries to come. The Kalmar Union dissolved in 1523.

Major conflict between its former components, Sweden-Finland and Denmark-Norway, ensued. During the Thirty Years' War (1618–1648), there were a series of battles fought between the contending sides. Many of these took place on what is today German territory. The wars were fought for different reasons, e.g. religious, economic and/or political. Sweden emerged as a major European power during this time. Much of what are today the Baltic countries, as well as parts of northern Germany and Poland, fell under Swedish rule. Its enhanced position as a great power in the Nordic region and in Europe was further strengthened when Denmark was forced to cede to Sweden control of the provinces of Skåne, Blekinge and Halland in 1658. The beginnings of the present-day boundaries of the Nordic countries were then discernible from that time forward—though Finland continued as part of the Swedish realm and Norway continued to be attached to Denmark. A subsequent series of wars with Poland and Russia eventually eroded Swedish dominions in the Baltic area. Finland was lost to Russia in 1809 but became an independent nation in 1917. Following the Napoleonic wars, Norway was separated from Denmark and became part of a political union with Sweden that continued until 1905, after which Norway too secured its own independence.

The political status of Iceland, the Faeroe Islands and Åland took additional time to solidify. Iceland over several centuries was treated as a colonial dependency first by Norway and then Denmark. It was not until 1944 that the island received its full independence. Greenland's colonial link to Denmark continued still longer—receiving full autonomy in domestic affairs only in 2009 while remaining dependent on Copenhagen in its foreign and defense relations. The Faeroe Islands, also long linked to Denmark, achieved a similar autonomous status under the Danish crown in 2005. Interestingly, while Denmark joined the European Union in 1973 both the Faeroe Islands and Greenland have remained outside. The unique status of Åland—historically linked to both Sweden and Finland—was not finally settled until 1922. This small community (predominantly Swedish-speaking but politically part of Finland) was also given an autonomous status.

The last portion of the nineteenth century and the first part of the twentieth were times when not only the political boundaries of the Nordic region were solidified, but also a period when the main features of their modern societies began to take shape. During this era, Nordic societies began the transformation from being rather poor, rural and traditional societies into becoming the prosperous, increasingly urban and sophisticated societies of today. Industrialization and the expansion of educational and training opportunities combined to create a growing and skilled labor force that brought new prosperity and confidence to the region. Most of the countries of the Nordic region were able to avoid the devastation of the First World War. Democratic reforms were instituted leading to the expansion of the franchise and to popularly elected responsible governments. It was a period when labor unions became quite active and influential, and the first social democratic parties entered the political arena and received wide support. It is suggested that roots of the modern-day social welfare states in the Nordic countries were first established during this era (See Chaps. 2 and 4 for additional details).

The coming of the world-wide depression of the 1930s and the onslaught of the Second World War temporarily brought a halt to this forward progress. The latter conflict proved to be a major trauma for Norway and Denmark who were occupied by Nazi Germany. Finland suffered from its own military engagements with the Soviet Union. Iceland came under the wartime supervision of the Allies. Only Sweden was able to maintain its sovereignty and neutrality throughout the war. There were significant repercussions which arose from this era (Refer to Chap. 7 for further information). At the war's end, each of the Nordic counties looked for opportunities for economic and societal security and renewal (See Chap. 5 for additional details). They pursued such goals through a variety of means and structures including greater cooperation among themselves. Much of their initial efforts had to take place within the context of an emerging Cold War between the United States and the Soviet Union. The resulting forms of Nordic cooperation became a distinctive hallmark of the societies that developed from that point forward.

1.3 The Development of Nordic Cooperation After World War II

The development of the Nordic Region within an overall European setting can be traced to the intensified cultural, political and policy cooperation which emerged in the aftermath of World War II. As noted above, Norway, Finland, and Denmark, were heavily affected by that conflict. When wartime was over, discussions regarding Nordic defense cooperation were initiated among these countries including Sweden, and proposals were made to organize a Nordic Defense Union (Wahlbäck, 2000). However, their different experiences during the war, as well as divergent expectations and judgements regarding future risks, made it impossible to accomplish this. Nonetheless, the shared cultural traditions and good relations among people of the different Nordic states did establish a foundation for a further strengthening of Nordic cooperation in other dimensions.

Nils Andrén (2000) has provided an overview of the evolution of "cultural nordism". He stresses the role played by "Föreningen Norden," a civil society organization which was established in all five countries after the end of World War I and reemerged as a powerful uniting force after World War II. The influence of its work and other similar bodies can be seen in the actions of the Nordic Ministers of Education, who were also responsible for cultural cooperation. In 1946, they established a forum called The Nordic Cultural Commission, which gave such organizations as the "Föreningen Norden" the right to participate in meetings and to provide ministerial staff with insights and experiences. This type of organization, as well as many other networks and civil society activities across the national borders, further strengthened the shared culture and good relations among the countries of the Nordic region. Jan Andersson (2000) convincingly argues that these efforts contributed to the creation of the Nordic Council (NC) in 1952 and to its successful work during years to come.[3] The NC became the official body for inter-Nordic cooperation and the development of common initiatives among the five Nordic counties and the three autonomous communities.

The establishment of the NC coincided with a period of economic growth in the Nordic region which laid the ground for the development of an effective labor market and social policies in the postwar era (Refer to Chaps. 2, 4 and 5 for additional information). Both were high on the political agendas of all the Nordic countries. In the process of sharing ideas and proposals, the NC to contributed to making the region internationally known for its innovative and progressive welfare policies. Also, an additional priority of the NC was its free movement initiative which resulted in a decision to abolish the requirement of passports for Nordic citizens within the region. This was a means to further deepening both the economic and cultural cooperation in the region. It was temporarily abandoned in 2015 due to an extremely heavy influx of migrants and later, also again temporarily, because of the Covid-19 pandemic (Ljunggren, 2022).

Attempts were also made to formulate an overall economic plan for the Nordic region, focusing on the establishment of a Nordic customs union and a Nordic financing fund.[4] The main reason for doing so was to increase economic interdependence between the Nordic countries and other European countries and to facilitate joint Nordic action during trade negotiations with the European Economic Community (EEC).[5] The content of such an economic plan for the Nordic countries was discussed by the NC in 1968 and a decision was taken to formulate such a scheme. However, after years of discussion within each of the five countries, and negotiations between them regarding the specific content of the plan, it became obvious that such an agreement could not be reached. Therefore, in 1971, the NC decided to move forward with sector-by-sector agreements in various key policy areas. In conjunction with this decision, the Nordic Council of Ministers (NCM) was established and, together with NC, it was given the main responsibility for implementation of sector-based Nordic policy cooperation.[6]

---

[3] As an inter-parliamentary organization, the NC has 87 members from Denmark, Finland, Iceland, Norway, Sweden, the Faroe Islands, Greenland and Åland.

[4] This paragraph builds on Wiklund (2000a; 2000b).

[5] At this point in time France, Germany, Italy, Belgium, Netherlands, and Luxemburg were members of EEC, the goal of which was to abolish customs barriers between member countries. Thereafter EEC expanded considerably but ceased to exist in 1993, when the European Monetary Union (ERC) was formed.

[6] The Nordic Council of Ministers (NCM) is the official body for inter-governmental cooperation in the Nordic region. The Presidency of NCM rotates yearly among the five countries. The prime ministers take collective responsibility for the work of the body and allocate decision-making

The Nordic economies began to recover, and then rapidly accelerated during the two decades following the Second World War. Such strong economic growth enabled them to fund relatively generous welfare systems in each of the Nordic countries. During this period, both formal and informal cooperation between the Nordic neighbors continued to strengthen. The establishment of the NCM initiated an ongoing process of incremental policy coordination across several sectors including matters related to defense, education, and health. This type of incremental regional cooperation became a hallmark of public policy development in many areas including labor practices and support for scientific research (Sundelius & Wiklund, 2017a, 2017b). Other efforts included the creation of a Nordic Investment Bank (Wiklund, 2000a, 2000b, 2000c), the initiation of common labor market requirements, and the establishment of the Nordic Industrial Foundation.[7] As Enestam (2017) has suggested, such examples of informal cooperation can be as important to Nordic regional development as more formal agreements on policy.

1.4 Current Challenges Facing the Nordic Region

Over the most recent decades, the influences of globalization and liberalization have affected both political debate and public policy development in the Nordic region. They have had a major impact on public attitudes and opinion regarding the continuation of the Nordic welfare state and the various services it has traditionally provided. Such discussions have forced the Nordic community to consider how public policy is best developed and implemented and the manner in which it can be updated and financed. These challenges have been addressed within each of the individual countries and within regional fora. One can identify both common and divergent policy approaches. These are the focus of a number of the chapters in this volume.

An equally important trend according to Sundelius and Wiklund (2017a, 2017b), is the growing focus accorded defense and civil security policies in the region since the start of the current century (See Chap. 7 for additional details). The increased attention given to defense policy was manifested in the establishment in 2009 of the Nordic Defense Cooperation (NORDEFCO). The goal of this effort was to facilitate common solutions and strengthen the defense capabilities of the Nordic countries (Westberg, 2017). More recently, the unexpected war in Ukraine, has resulted in Finland and Sweden joining the North Atlantic Treaty Organization (NATO). This is an equally clear sign of the increased importance of defense cooperation in the Nordic region.[8] The emergence of the Arctic as a potential zone for both cooperation

---

responsibility to their Ministers for Nordic Cooperation (MR-SAM). The day-to-day work is carried out by a Committee for Cooperation (NSK).

[7] These and other examples are listed in the Caldarium attached to Sundelius and Wiklund (eds) (2000).

[8] Denmark, Norway and Iceland became NATO-members already in 1949.

and conflict has also become more of a concern to the Nordics (See Chap. 8 of the book for further details).

Civil security policies have also been given higher priority during this period by the individual Nordic states and collectively through Nordic cooperation (Bailes & Sandö, 2017) Starting in 2009, high-level representatives from the five Nordic governments met at the Haga Castle in Sweden with the purpose of strengthening cooperation and preparation for extraordinary civil crises such as terrorism, pandemics, or natural disasters. Thus, the so-called "Haga process" was initiated. Its objective was to formulate and implement a comprehensive Nordic plan for crisis preparedness, with clear priorities including prescriptions for joint Nordic action on international arenas and platforms.

Significantly, additional steps regarding civil preparedness and civil society cooperation were still absent when, in 2015, the war in Syria caused a wave of migration to the Nordic countries. Shortcomings in the joint Nordic preparedness structures rapidly led to a full stop in the established freedom of movement across national boundaries in the region. Although implemented only for a limited amount of time, the event represented a resounding retreat from the decades-long principle of freedom of movement among the Nordic peoples. Equally portentous, the failure of Nordic cooperation to prevent the Nordic countries from cordoning off their national boundaries under the Covid-19 pandemic still resounds in popular and political debates within the region.

However, a joint commitment to cooperate on security across national borders in the Nordic region seems to have survived the hiccups of the migration and pandemic crises. This was confirmed at the Nordic prime ministers meeting of August 2022. There it was stressed that joint efforts to deepen the dialogue on security policy should continue as before (Bjarnsson, 2022).

Another policy field given high priority in terms of the Nordic cooperation agenda relates to the challenges and consequences of climate change for the environment. Even though this policy area has long been given attention by the Nordics, it has now become even more urgent in their eyes. As climate change brings new economic, health and social challenges to the region, the Nordics have sought to address these problems collectively (As an example, see Chap. 9 of this volume). They have been strong supporters of the Paris Climate Agreement and have acted in accord with UN's seventeen Sustainable Development Goals (SDG).[9]

Quite a few of the Nordic policy initiatives that have been undertaken in these areas have been designed to address the negative consequences of climate change and to secure more knowledge and better understanding of the processes involved. A book that influenced the Nordic debate is *Hot Topic—Cold Comfort: Climate Change and Attitude Change* (2012), written by the Norwegian Professor of Sociology and former Minister of Higher Education and Research Gudmund Hernes. He argues that there have been several events during the last hundred years which,

[9] The Paris agreement is a legally binding international treaty on climate change adopted in 2015 and the seventeen SDG- goals are part of the 2030 Agenda for Sustainable Development approved of at the United Nations sustainable Development Summit in 2015.

when taken together, are creating what he calls an "ecological revolution" in our thinking about the interaction between human beings and nature. Hernes suggests that we in the past perceived the Earth as invincible but now must realize that it is fragile. Therefore, he believes that politicians now must be willing to think along new lines. One example was the decision by the Nordic prime ministers in 2008 to fund joint research collaboration on climate, energy, and the environment. This so-called "Top-level Research Initiative"[10] was followed by another large research program sponsored by NordForsk that was focused on the Nordic Arctic. The goals and research results stemming for the latter effort have been published in the book *Nordic Perspectives on Responsible Development of the Artic: Pathways to Action* (Nord, 2021).

In the coming years it is likely that the Nordics will continue to prioritize each of the policy areas noted above and to engage in further cooperative efforts among themselves. While the ongoing processes of globalization have resulted in increasing amounts of policy learning across regions, the NC, NCM and other Nordic institutions are most probably going to continue to provide important avenues for regional inspiration and collaboration. As such, the process of identifying the extent to which policy choices in the Nordic region are reflective of global, regional or individual state priorities and concerns will be an ongoing task.

1.5 The Elusive Nordic Model

When discussing the nature of public policy in the Nordic region, reference is often made to a supposed "Nordic Model." Some historians seem to agree that the concept of a Nordic Model springs from a kind of interaction between national and international political agendas, sometimes inspired by political debates and usages far away from the Nordic region, and sometimes within it. There have also been times when an emphasis was placed on a Danish or Swedish Model, partly to attract attention to particular success stories in each country (Hellenes, 2022). Since the 1980s, however, the dominant reference has been to a Nordic Model (Stråth, 1993). The concept is frequently utilized by analysts as both a touchstone and compass to explain the essential character and direction of both domestic and foreign policies pursued by the nations of this area. Yet despite its regular invocation, the precise features and components of this model are rarely fully detailed. An all too easy assumption is readily made that historic links, shared values and cultural affinities between the Nordic societies tend to bring forth a uniform set of policy options and perspectives that are broadly adopted throughout region. The Nordic Model continues to operate often as a cherished but largely elusive idea.

The elusive character of the "Nordic Model" may stem, in part, from its multifunctional application. On the one hand, it is often used as a way of denoting a distinctive

[10] Detailed information regarding the "Top-level Research Initiative" is found in the report Solving the Climate Crises – a Nordic Contribution (2015).

form of public policy found only in the northern reaches of Europe. In this regard, it helps to promote the idea of a unique regional public policy that is responsive to the specific needs and aspirations of the communities found within this area. Nordic public policy is said to be "exceptional" because it is unlike that found in other places of the world. It is claimed that such policy has a history of respect for individual liberty combined with a peasant community tradition. It is seen to have produced a distinctive system of public welfare, a "middle way" between capitalism and communism and, at times, some of the most equal societies in the world. In contrast to the rest of Europe, the Nordic countries have been regarded as "non-Catholic, non-colonial and non-imperial" (Hilson, 2008: 16).

On the other hand, the "Nordic Model" can also be invoked by other observers as an aspirational concept rather than simply a descriptive summation of policy formulation and implementation within a particular region. Here it is utilized to promote the idea that other societies might follow the Nordic example and emulate the policy decisions and actions taken there and apply them to their own situations. Whether in discussions about education, health care or the environment, the "Nordic Model" is frequently used in such a heuristic fashion. Here, Nordic public policy is "exceptional" in the sense that it is deemed meritorious and an example that should be followed by others.

Yet, does the reality of today's world suggest that the Nordic societies are broadly "exceptional" in either of these two usages? Are current public policies pursued by the countries of the Nordic region that different from those found in other post-industrial societies around the globe? Politically, the Nordic countries seem to be on a similar path with other European and North American societies experiencing major debates and policy revisions concerning migration, public values, and cultural issues related to identity. As small and free-trading nations, the Nordics are subject to the impact of globalization, interest rate instabilities, and energy market uncertainties just like many other societies. Perhaps the policy challenges and choices confronting the Nordics states today are not that dissimilar from other communities. How different will be the policy options that they will pursue?

Also, it is perhaps less clear than in the past whether the Nordic Model is to be emulated. Recent initiatives in the field of "green" re-industrialization, based on investments in CO_2-free steel production and modern car battery manufacturing seem to hold some promise of continued Nordic leadership, innovation and economic success. However, such undertakings have also been accompanied by land and resource use disputes (particularly with the region's Indigenous peoples) that has made the whole process of finding a new path to resource utilization less clear and more complicated (See Chap. 10 of this book for details). Green energy production in the form of wind turbine parks have also given rise to public dissatisfaction and sparked protests. Such examples of unease over recent Nordic public policy choices raises questions as to whether they are worthy of emulation and adoption by others?

Likewise, a good case can be made for saying that there may be greater divergence among the Nordic states in their public policy choices than is popularly imagined or is promoted by some in the idea of a *Nordic Model*. This has been suggested recently by several scholars. They note that while "The Nordic countries have influenced each other's policies and have learned from each other's experiences" they have also "been free to apply bespoke national solutions when it suited them." (Koivunen et al., 2021: 5). Perhaps there is a greater degree of inner diversity within the region with respect to public policy formulation and implementation than has been traditionally portrayed.

To what extent does a Nordic Model really exist? Are we looking at somewhat similar, but quite distinctive, national models and approaches to economic, social, and environmental challenges? Are we simply talking about a concept which stirs political imagination and parlance but remains open to a variety of difficult questions? The purpose of this book is to provide some empirical evidence regarding changes in Nordic public policy over time as well as an assessment of current degrees of exceptionalism and inner diversity found therein. The following essays focus on policy areas of crucial importance to the region and consider whether a *Nordic Model* can be identified and defined. Our ultimate goal is to lay the foundation for a more nuanced understanding of the Nordic region and policymaking within it.

References

Andersson, J. A. (2000). 1950-talet. Tid att så—tid att skörda. In B. Sundelius, & C. Wiklund (Eds.), *Norden i sicksack—Tre spårbyten inom nordiskt samarbete* (pp. 67–90). Santérus Förlag.

Andrén, N. (2000). Nordiska kulturkommissionen lägger grunden. In B. Sundelius, & C. Wiklund (Eds.), *Norden i sicksack—Tre spårbyten inom nordiskt samarbete* (pp. 47–66). Santérus Förlag.

Bailes, A. J. K., & Sandö, C. (2017). Nordiskt krisberedskapssamarbete Haga- processen 2009–2015. In B. Sundelius, & C. Wiklund (Eds.), *Norden sett inifrån—Det fjärde spårbytet* (pp. 57–74). ScandBook, Santérus Förlag.

Baldersheim, H., & Stålberg, K. (Eds.). (1994). *Towards the self-regulating municipality—Free communes and administrative modernizatrion in Scandinavia.* Ashgate Publishing Limited.

Bjarnsson, B. (2022). Nye Muligheter i Nordiskt samarbete—Fred, frihed og demokrati. *Nordisk Tidskrift, 4*(22), 409–414.

Enestam, J.-E. (2017). Nordiskt och svenskt-finskt försvarssamarbete. In B. Sundelius & C. Wiklund (Eds.), *Norden sett inifrån—Det fjärde spårbytet*. ScandBook, Santérus Förlag.

Hellenes, A. M. (2022). Tracing the Nordic Model: French creations, Swedish appropriations and Nordic articulations. In L. Mjöset, H. Byrkjeflot, M. Mordhorst, & K. Petersen (Eds.), *The making and circulation of Nordic models, ideas and images.* Routledge.

Hernes, G. (2012). *Hot topic—Cold comfort. Climate change and attitude change.*

Hilson, M. (2008). *The Nordic model: Scandinavia since 1945.* Reaktion Books.

Hood, C. (1991). A public management for all seasons? *Public Administration, 69*(1), 3–19.

Koivunen, A., Ojala, J., & Holmén, J. (2021). *The Nordic economic, social and political model: Challenges in the 21st century.* Routledge.

Ljunggren, A. (2022). Krönika om Nordiskt Samarbete—Gränskontrollerna i Norden. *Nordisk Tidskrift, 3*(22), 367–370.

Ljunggren, A. (2023). Högervindarnaoch den Nordiska Demokratin. *Nordisk Tidskrift, 2*(23), 215–219.

Lugmayer, H. (2009). *Alltingret på Tingvalla.* Iceland Review.

Minority Rights Group. *Sami in Finland.* Retrieved June 2, 2025, from https://minorityrights.org/communities/sami/

Nordic Council of Ministers. *Fakta om de nordiska länderna.* Retrieved June 2, 2025, from https://www.norden.org/sv/information/fakta-om-de-nordiska-landerna

Nordiskt samarbete. (2025). *Fakta om de nordiska länderna/ Facts abort the Nordic countries, accessible in English at* https://www.minorityrights.org/communities/sami

Nordnorge.com. *The Sami people—Indigenous people of the north.* Retrieved June 2, 2025, from https://www.nordnorge.com/en/topic/the-sami-are-the-indigenous-people-of-the-north/

Pollitt, C., & Bouckaert, G. (2000). *Public management reform: A comparative analysis.* Oxford University Press.

Solving the climate crisis: A Nordic contribution. (2015). (1 ed). NordForsk.

Stråth, B. (1993). Den nordiska modellen: Historisk bakgrund och hur talet om en nordisk modell uppstod. *Nordisk Tidskrift, 69*, 55–61.

Sundelius, B., & Wiklund, C. (2000). *Norden i sicksack—Tre spårbyten inom nordiskt samarbete.* ScandBook, Santérus Förlag.

Sundelius, B., & Wiklund, C. (2017a). *Norden sett inifrån—Det fjärde spårbytet.* ScandBook, Santérus Förlag

Sundelius, B., & Wiklund, C. (2017b). Que vadis? Tretton insikter om Norden. In B. Sundelius, & C. Wiklund (Eds.), *Norden sett inifrån—Det fjärde spårbytet* (pp. 311–333). ScandBook, Santérus Förlag.

Truth Commission for the Sami People. *About the indigenous Sami.* Retrieved June 2, 2025, from https://www.sanningskommissionensamer.se/en/about-the-indigenous-sami/

Wahlbäck, K. (2000). Förbubdsförhandlingarna 1948–49. In B. Sundelius, & C. Wiklund (Eds.), *Norden i sicksack—Tre spårbyten inom nordiskt samarbete* (pp. 31–45). ScandBook, Santérus Förlag.

Westberg, J. (2017). Försvarssamarbetet en ny väg till fördjupat nordiskt samarbete. In B. Sundelius, & C. Wiklund (Eds.), *Norden sett inifrån—Det fjärde spårbytet* (pp. 41–55). ScandBook, Santérus Förlag.

Wiklund, C. (2000a). Femte gången gillt—Nordiska investerinsbankens tillkomst. In B. Sundelius, & C. Wiklund (Eds.), *Norden i sicksack—Tre spårbyten inom nordiskt samarbete* (pp. 145–182). ScandBook, Santérus Förlag.

Wiklund, C. (2000b). Från helhetsplaner till sektorssamverkan. In B. Sundelius, & C. Wiklund (Eds.), *Norden i sicksack—Tre spårbyten inom nordiskt samarbete* (pp. 125–143). ScandBook, Santérus Förlag.

Wiklund, C. (2000c). Nordek-planen och dess föregångare. In B. Sundelius, & C. Wiklund (Eds.), *Norden i sicksack—Tre spårbyten inom nordiskt samarbete.* ScandBook, Santérus Förlag.

Niklas Eklund is Professor of Politics and Public Administration at the Department of Political Science, Umeå University, and the current Director of the European CBRNE Center. He also serves on the steering committee for the Arctic Centre at Umeå University and on the board of the local twenty-first Century Scandinavian Policy Studies Foundation. In his research, he specializes among other things in Arctic security, geopolitics and Russian studies. He was appointed Arctic Six Chair in Nordic and Arctic Security for 2024–2026, a position from which he is actively trying to promote transboundary Nordic research initiatives driven by geopolitical change and emergent security needs in the European High North.

Gunnel Gustafsson is Professor Emerita in Political Science at Umeå University. She served as Director of NordForsk from 2010–2017. From 2005–2010 she was Deputy Director General of the Swedish Research Council and prior to then, Pro Vice Chancellor of Umeå University. She has also served on several editorial boards, peer research review committees and contributed to the work of a number of research financing agencies including the Bank of Sweden's Tercentenary Foundation. She has written widely on topics related to Swedish and Nordic public policy with a special focus on education, social welfare and democracy. She has also led research efforts aimed at studying local government reform in Sweden, women's political leadership in that country, and Nordic efforts to secure gender equality.

Chapter 2
The Evolution of Public Policies in the Nordic Countries

Anders Lidström

Abstract This chapter outlines a perspective on policy change in the Nordic countries after the Second World War. It is argued that three distinctive periods can be identified. The first is a tax financed expansion period of social welfare and public services during the first three decades after the war. When tax increases ceased to be functional, a liberalization and globalization phase took over. It was hoped that more competitive businesses would generate a surplus that would be sufficient for retaining the public commitments, including the welfare sector. Markets were deregulated, previous state-owned businesses were sold off and private providers were invited to carry out welfare services. New Public Management became the leading philosophy of how the public sector should be governed. The third phase, a post-global era, emerged around 2020 with increasing awareness of on-going climate change, the Covid-19 pandemic, the Russian invasion of Ukraine, and the retreat from an international rules-based world order. Under such conditions, the Nordic countries' reliance on liberalization of trade became a very risky strategy and was not a safe basis for generating public resources. Also, additional demands for public services emerged. New infrastructure was needed to handle floods, forest fires and droughts. Resources had to be stored for unseen events. The national defence needed to be strengthened, and Finland and Sweden joined the NATO. A new approach had to be considered.

Keywords Public policy · Nordic countries · Welfare state · Liberalization · Globalization · Post-global era

2.1 Introduction

The character, conduct and financing of public policies in the Nordic countries have undergone significant changes during the past several decades. This chapter outlines a perspective on how these changes can be understood. Although several distinctive

A. Lidström (✉)
Umeå University, Umeå, Sweden
e-mail: anders.lidstrom@umu.se

D. C. Nord (ed.), *Contemporary Nordic Public Policies*, Springer Polar Sciences,
https://doi.org/10.1007/978-3-032-11292-7_2

traits of the Nordic countries such as strong trade unions, peaceful relations between labour and capital and a pragmatic decision-making culture were in evidence prior to the Second World War, it was not until after the end of that conflict that the role of government started to expand. The extent of this expansion eventually made these countries stand out compared to the rest of the Western world. Globally, war-time planning gave rise to the view that the state could play a more active role, but nowhere did this approach to public policy reach such a high level as in the Nordic countries (Tanzi, 2011).

The Nordic public welfare systems were to become internationally noteworthy in several ways. Not only were the levels of benefits and services exceptionally generous, and the level of taxes that citizens paid higher than anywhere else, but the content of the welfare system also became largely based on the principle of universality, rather than means-testing (Castles, 1993; Esping-Andersen, 1990; Kautto et al., 2001). This type of system could probably only have emerged in the Nordic societies as they are characterized by a high level of trust between citizens and in public authorities, along with a low degree of public corruption. Paying high taxes requires that citizens can rely on the reality that in a vulnerable situation, they will get the help they need on an equal basis and that public resources will be used in an efficient and effective way (Rothstein, 2000).

However, over the subsequent decades, this extensive role by government in the Nordic countries has undergone significant changes. Two distinctive phases can be identified and in the early 2020s, the systems seem to have entered a third and entirely new, post-global phase.

2.2 The Tax Financed Expansion Period of Social Welfare and Public Services in the Nordic Countries

The government sphere of activity in the Nordic countries grew rapidly during the 30–40 years following the Second World War, and much faster than in other Western countries. This was in marked contrast to the region's previous experience. For example, the public expenditures of both Norway and Sweden as a share of their respective GDPs were considerably below the average of other countries who would later form the OECD. However, by 1960, they were at the same level as these other countries, and by 1990, much higher (Tanzi & Schuknecht, 2000). By 1995, public spending as share of the GDP had increased to exceptionally high levels throughout the Nordic region. For example, it rose to 65.2% in Sweden, 61.5% in Finland, and 59.3% in Denmark. The situation was much more modest in the other OECD countries, such as the UK (43.9%), Germany (48.3%), the US (37.1%) and Japan (36.7%) (Tanzi, 2011). In Denmark, Norway, and Finland, this expansion took place despite the pressures of recovering from occupation during the Second World War, damaged infrastructure, and, in the case of Finland, having to pay extensive war reparations to the USSR.

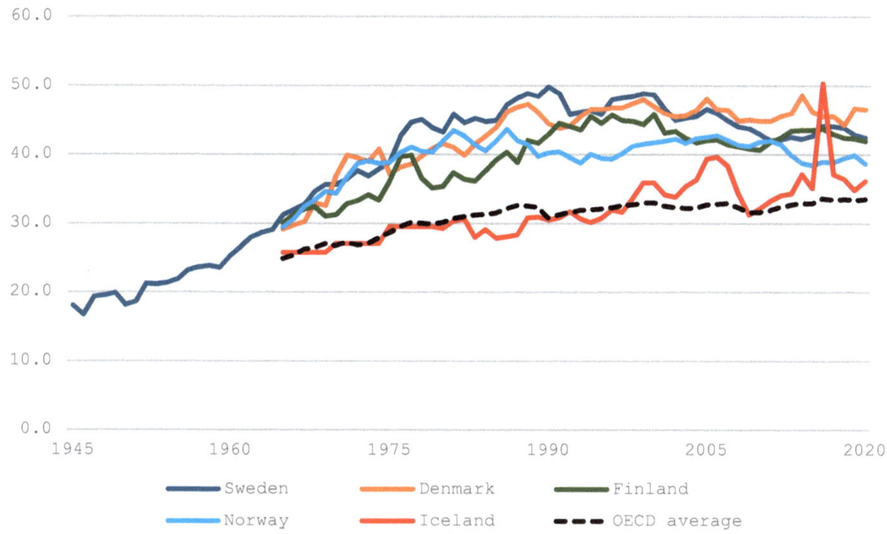

Fig. 2.1 Tax share of GDP for Sweden and the other Nordic countries

This expansion of the Nordic social welfare state was financed by taxes, first mainly derived from personal income taxes, but later complemented by value added taxes. This was possible to bring about due to three decades of extensive economic growth in the region. It was also facilitated by strong national and Social Democratic governments with a clear purpose to improve the living conditions of ordinary citizens. They did not hesitate to use the tax instrument to achieve this aim. The result was a redistribution of wealth, which consequently gave the Nordic countries the most equal distribution of economic resources in modern history. They could point to Gini coefficients close to 0.20 in the middle of the 1980s (OECD, 2015). In Sweden, the tax share of the GDP increased from 18 to 50% from 1945 to 1990. The growth rates were similar in the other Nordic countries (See Fig. 2.1).

During most of this period, the Social Democrats were in power across the region. However, the role played by this party in the expansion varied in scope—from being crucial in Sweden to being more limited in Finland, where reforms generally required broad compromises among most of the political parties (Sipilä et al., 2009).[1] The expansion policies were also accepted, in principle, by most of the other political parties in all of the Nordic countries.

The Nordic welfare state also emerged as a local and regional phenomenon. Although the welfare systems emphasized national coherence and an equal treatment of all citizens, regardless of where they lived, local government was entrusted with implementing most of the welfare services regarding health care, education,

[1] During the 35-year period 1946–1980 Social Democrats were in government (on their own or as the leading party in a coalition) for 31 years in Sweden, 28 years in Norway, 26 years in Denmark and 24 years in Finland).

and social services (Sellers et al., 2020). To ensure that all municipalities had the capacity to carry out these new responsibilities, local governments were consolidated into larger units. In Sweden and Denmark, it was done on a compulsory basis. In Norway, Iceland, and Finland it was accomplished mainly through voluntary agreements (Hesse, 1991). At the regional level, a standard model soon emerged in Denmark, Sweden, and Norway, which featured directly elected county councils having responsibility for hospitals, and in Sweden, also health and dental services (Torfing et al., 2015).

The Nordic welfare state expansion also generated a rapid increase in the nursing professions, that mainly attracted women. While this contributed to a highly gender skewed labour market, it also made it possible for many women to get salaried employment and a wage of their own (Castles, 1998). In contrast to many other countries, the women's movements in the Nordic countries favoured transferring welfare responsibility from the voluntary sector to the state (Hernes, 1987). High female participation rates in the workforce became a further distinctive characteristic of the Nordic countries.

Throughout this period of expansion, Keynesianism was the dominant economic doctrine, emphasizing the use of government spending for counter-cyclical purposes. The Nordic economies were also primarily internally focused. As such, their national economic and tax policies could be pursued largely without fear of major international trade or investment consequences (see Chap. 5 in this volume for further discussion).

During this period the focus on the nation state was also on defence and security concerns. After the Second World War, Norway and Denmark joined NATO, but Finland and Sweden retained their policy of neutrality. This required the latter states to undertake extensive spending on national defence but also a need to maintain standby stocks of supplies in case that the countries were isolated or subjected to military aggression. Such stores encompassed both military and civil items including food. A self-sufficient agricultural sector that could support the population in times of crisis was also prioritized (see Chap. 7 in this volume for further discussion).

Although some assumed that this positive economic situation would continue forever (even allowing for the very optimistic pensions reforms of Sweden in the 1960s) evidence soon started to accumulate that suggested the opposite. Early indicators of a coming economic crisis appeared already at the beginning of the 1970s. The international oil crisis of that era reminded all that the extensive Nordic growth period had been fuelled, in part, by access to cheap energy. It also demonstrated how vulnerable and dependent national economies had become on free trade with the rest of the world. This was certainly the case of the Nordics. Hence, purely nationally based solutions became less viable. In the 1970s and 1980s, growth rates in the Nordic countries slowed down and soon became lower than in most other West European countries (Tanzi, 2011). Some critics also claimed that the Nordic states now suffered from ungovernability and government overload (Rose, 1984). Economists and non-socialist parties argued that taxes had reached a level that was harmful for society by reducing incentives for economic development. As this argument was eventually accepted even by the Social Democratic parties, the tax-based growth strategy came to an end.

In Sweden, a major tax reform was implemented in 1991 that considerably reduced marginal tax rates for the highest income earners and increased value-added taxes. Except for the Left Party, there was full political consensus on the question of ending tax increases. Another significant reform was to tax capital gains separately, instead as previously, adding it on top of the taxable wage income. This reduced marginal tax rates considerably, but at the expense of income equality. Similar tax reforms were implemented in the other Nordic countries (Tanzi, 2011).

By the end of this period, the political dominance of the Social Democrats also came to an end. Apart from in Finland and Iceland, Social Democratic parties had been continuously in power throughout the region since the end of the war—in Norway until 1965 and in Denmark until 1968. In Sweden this shift came in the 1976 elections, ending 44 years of the Social Democratic governance. The new electoral pattern for the region became one of governments with different political orientations replacing each other at regular intervals. As existing public policies were only partly rescinded when a new majority came into power, national policymaking now became a mixture of policies of the political left and right. Parties like the Greens and a variety of populist parties soon emerged that represented new political perspectives.

2.3 The Liberalization and Globalization Phase of the Policy Development

As mentioned earlier, policies directed toward economic growth were primarily a national affair during the first phase of post-war development. Although the Nordic countries where highly dependent on international trade, they could develop specific national policies for economic development and set taxes without undue concern for their negative effects on national competitiveness. However, with their economies becoming more intertwined with the rest of the world, this became increasingly difficult. With further tax increases virtually impossible, a new strategy was required to safeguard the resources needed for the public sector, especially for retaining the welfare state. This became particularly obvious during the financial crisis that began in the 1990s, when tax cuts were regarded as necessary to reduce public debt and to stabilize the economy.

The new strategy that was pursued not only fully accepted that the Nordic countries were in an internationally competitive position but, also, exposed these countries even further by reducing credit regulation and trade obstacles (Kosonen, 2001). Deregulation made low-wage countries, particularly in East Asia, more competitive in the manufacturing the goods. This together with increased trade across the globe contributed to the de-industrialization of Western countries.

At this time, Sweden and Finland also decided to join the European Union (Denmark already being a member) which contributed to the new course of action. Greater free market capitalism and further liberalization was assumed to enhance

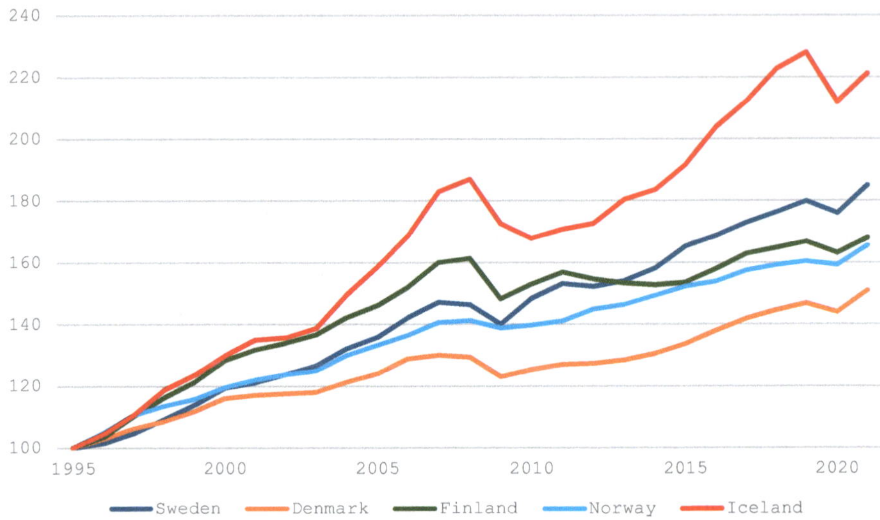

Fig. 2.2 The development of GDP in set prices for the Nordic countries 1995–2021

economic growth in a way that would generate more resources for the public sector, without necessitating any increases in taxes. This way of thinking was clearly inspired by the neo-liberalist attitudes that had developed in the UK and USA during the 1980s and were adopted by the Thatcher and Reagan governments. This new approach replaced Keynesianism as the predominant ideology of capitalism.

Indeed, this became a world-wide phenomenon, affecting not only the Western economies but also those of the Eastern Europe, Africa and Latin America as the decade proceeded (Palley, 2005). This approach was also strongly enhanced by increased digitalization of society, which facilitated the spread of information on a global scale. Despite the liberalization approach being more closely related to non-socialist ideological preferences, it also adopted by the Nordic Social Democratic governments as an alternative way of providing resources for their welfare systems. The strategy turned out to be successful and the promised economic growth was largely achieved. From 1995 to 2021, the GDP of the Nordic countries increased between 150 and 220% in real terms (see Fig. 2.2). Nevertheless, in all these countries the period was also characterized by a weakening of the Social Democratic influence in terms of electoral support as well a reduction in their number of years in government.[2]

Also, as a consequence, the dramatic increases in welfare spending across the Nordic region came to a halt. From approximately the middle of the 1990s, spending on public welfare as share of GDP became more stable at about 20 to 30% (significantly lower in Iceland) (see Fig. 2.3). This was an obvious result of the end of the

[2] During the 35-year period 1981–2015 Social Democrats were in government (on their own or as the leading party in a coalition) for 22 years in Sweden, 20 years in Norway and 14 years in Denmark and Finland).

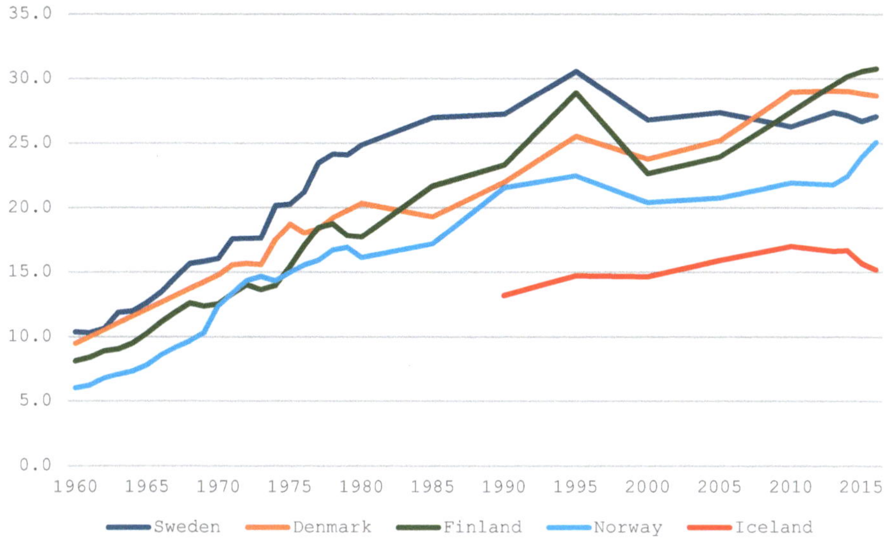

Fig. 2.3 Public expenditure on social policies as percentage of GDP for the Nordic countries 1960–2015

tax increases (as illustrated in Fig. 2.1) along with other reforms aimed at improving the efficiency of the public sector.

Another defining feature of Nordic public policies during this period were their growing reliance on local and regional government. More functions were transferred to municipalities and county councils, and they were given more independence to decide over their finances and services (Baldersheim & Ståhlberg, 1994; Lidström, 2007). This was accompanied by a shift in regional policy, now emphasizing that each municipality and county council was responsible for developing its own assets. This resulted in competition among them for citizens and businesses. This new approach replaced the more redistributive regional policies that had emerged during the tax financed period and which had provided central government assistance to parts of each country which faced the least favourable conditions for development (Hörnström, 2013).

The liberalization strategy, and a greater reliance on the market, had the intended policy effects. New economic growth generated the necessary resources to retain and even further develop public welfare in the Nordic region without tax increases. Figure 2.4 captures the situation in Sweden, showing how public spending in real terms continued to increase per head of population, despite relatively stable (and even reduced) taxes.

Until the early 1990s, when the era of annual tax increases ended (see Fig. 2.1), both private and public consumption levels grew at a fairly even rate. After 1992, when the new liberalization policies began to take effect, public consumption continued to increase, albeit at a slower pace. Hence, the intended trickle-down effect on public welfare was achieved, at least until around 2016. After that, the strategy seemed not

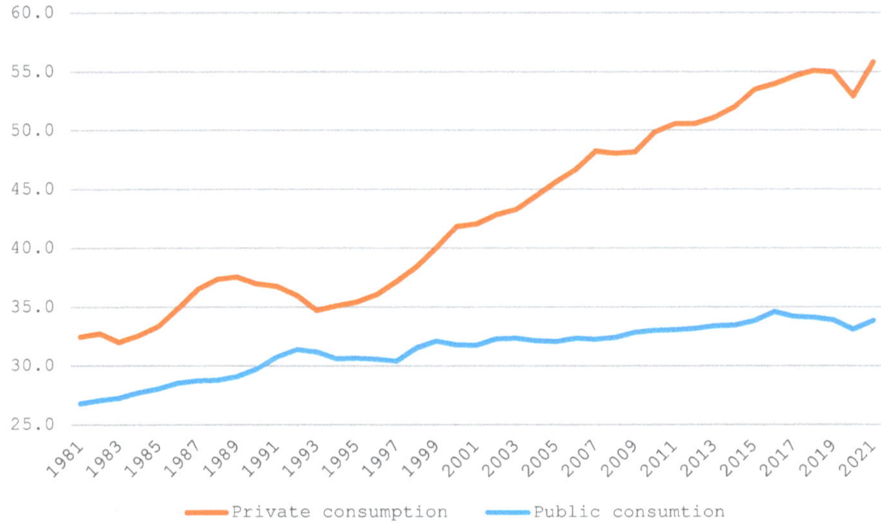

Fig. 2.4 Public and private consumption in Sweden 1981–2021, in set prices

to generate additional resources, but this may also have been affected by further tax reductions. However, as also seen in Fig. 2.4, the real winners of the liberalization strategy were private households. Private spending increased much faster than public consumption between 1993 and 2021. Some of this was undertaken by households purchasing social insurance and welfare services in the growing private welfare sector. Its operation, of course, was not based on the principle of equality.

Although Nordic citizens generally experienced a higher standard of living during the liberalization phase, the better off sectors of society benefitted to a greater extent than those with low incomes and no capital. The Nordic countries retained their position as among the most equal societies in the world even during the liberalization period, but inequalities increased (OECD, 2015). This was reinforced by a more open labour market that accompanied liberalization and EU membership in the case of Denmark, Sweden and Finland. A large influx of refugees from war-torn countries, particularly in Sweden, also had an impact.

Another policy assumption of the liberalization period was that as more international economic exchange took place, and stronger dependencies between countries emerged, this would reduce the risk for armed conflicts. Wars might break supply chains and make trade more difficult between countries and therefore costly for belligerents. The ending of the Cold War with the breakdown of communism in Eastern Europe and the Soviet Union, further contributed to this understanding that globalized free capitalism would make wars less likely. Accordingly, it was claimed that the previous strong national defences could be dismantled, and standby stocks of supplies could be sold off (although Finland was an exception in this respect). This "peace dividend" would reduce the need to maintain a high level of taxation as existing resources could now be used for other purposes.

In Sweden, government reserve stocks of food were sold off between 1995 and 2002. Sweden also abolished its state monopoly on pharmacies in 2009. It did so without establishing a national supply of pharmaceuticals for unforeseen events. Furthermore, national self-sufficiency within the agricultural sector was no longer regarded as necessary.[3] It was assumed that everyday consumer goods, as well as technologically advanced products, could be produced wherever in the world it was cheapest and delivered just-in-time through a reliable and well-functioning system of international trade.

2.4 Public Policies in the Post-global Era: Challenges of Climate Change, Ukraine, and a Weaker Rules-Based International Order

Fundamental components of the liberalization and globalization strategy were effectively challenged in the early 2020s. This was first seen during the Covid-19 pandemic and later became more evident with the Russian aggression towards Ukraine. In addition, it became increasingly obvious that global climate change would have costly consequences, even in the Nordic societies, through flooding, forest fires and droughts. The Covid pandemic together with frequent occurrences of extreme weather conditions during this period revealed a general lack of preparation for unforeseen events across the Nordic region. In many localities, the infrastructure was not adequately adjusted to cope with the new climate situation and previously existing critical medical supply reserves had been sold off. In the never-ending hunt for efficiency under neo-liberalism, goods that were assumed unlikely to be used had been regarded as too expensive to maintain.

The Russian aggression towards Ukraine, first with the occupation of the Crimean Peninsula in 2014 and later with the full-scale attack on Ukraine in February 2022, dramatically challenged the established norms of the world order and respect for national borders. The reestablishment of authoritarianism in Russia further undermined the international reputation of that country. The war also interrupted the flow of cheap energy, fertilizers and wheat and generally hampered international trade. Digital information flows were disrupted. As this happened close to the Nordic countries, these societies found themselves in a much more vulnerable geopolitical situation (see Chap. 7 of this volume for further information). Likewise, the government of China's more aggressive attitude toward both domestic political opposition and in its threats towards Taiwan, further added to a more uncertain international situation.

[3] There are no official statistics on the amount of self-sufficiency of Swedish farming as this varies between different types of goods. However, estimations from the Federation of Swedish Farmers (LRF) suggests that self-sufficiency in the 1990s was 75% and is currently only 50%. In Finland, self-sufficiency is 80%. See https://www.lrf.se/politikochpaverkan/foretagarvillkor-och-konkurrenskraft/nationell-livsmedelsstrategi/sjalvforsorjning/.

The established tenets of liberalization and globalization have been further challenged with the advent of the Trump administration in the United States. During his first administration, and now during his second, Trump has increased of trade tariffs and waivered in his support of NATO and Ukraine, thus rejecting key components of the rules-based international order that has existed since the Second World War.

Taken together, these developments undermined several assumptions that had been taken for granted by the Nordic policymakers during the liberalization and globalization phase. It now appeared to them that they could no longer rely upon globalization and international dependence to stop wars. Furthermore, Russia could not be trusted. It was no longer possible to depend on smooth and uninterrupted deliveries through international trade. Flows of energy could easily be cut off. Vulnerability to cyberattacks meant that digitalization could become a threat. Indeed, the liberalization strategy which underlay much of Nordic public policy suddenly became very risky. Such a new reality would affect both consumer goods and specialized highly technology items. Missing just one key component could undermine the production of a technologically advanced good. Dependence on Russian supplies of gas, fertilizers, or wheat was not to be recommended. As these new threats were unlikely to be reversed in the foreseeable future, a new strategy to generate the necessary resources for the public sector was needed. The Nordic countries have now entered the post-global era of policy development.

However, any new strategy comes at a cost. Nordic public policies can no longer rely on being financed by gains produced through global economic exchange to the same extent as before. Dependence on international trade and energy flows need to be reduced. Economic development will have to be confined more to the capacities of each nation and on cooperation within Europe. This will probably generate a slower rate of growth in the overall Nordic region and less of a surplus for public sector spending, unless existing levels of taxation are altered.

Yet, the new situation also requires additional public commitments. A strategic response to the new global order was the recent decisions of Finland and Sweden to join NATO alongside their Nordic neighbors. However, such membership necessitates an increase in military spending to the now required five percent level of GDP, and possibly even more. Farming investments and supports will also need to be restored to a level closer to national self-sufficiency. Likewise, additional spending on infrastructure improvements to withstand climate change and to provide resources for various types of "just-in-case" events will also be necessary. Taken together, this will likely mean that there will be less resources available for spending on public welfare.

On top of all this, the Nordic welfare systems will face other additional challenges. The increasing number of older people in these societies is one of these. Although pension ages have been raised and the elderly are healthier than in the past, inevitably, more public resources will need to be devoted to social assistance and health care. There is also will be a need to recruit additional personnel to the public sector and retain them through higher wage and better working conditions (see Chap. 6 in this volume in this regard). Likewise new public resources may be required for other unforeseen tasks given a more insecure and volatile world situation.

A further consequence of this new situation is that the Nordic public will be expecting that their national governments take on a more active role in providing for their security (see Chap. 7 in this volume for additional detail). Contrary to the situation during the liberalization and globalization phase, when many of these functions were hived off to both EU and local government, central governments across the Nordic region are now expected to take a more explicit lead in this effort. This includes restoring standby stocks of supplies, spending more on national defence, and investing in new infrastructure to meet climate change emergencies.

It also requires a new relationship between the central and local governments within the Nordic region. Regional and local governments are increasingly seen as being not effective in responding to the new security challenges. As differences in abilities and capacities between metropolitan areas and the rural communities of the countries have become more evident, particularly in Sweden, voices have been raised asking for more centralized solutions. In Finland, provincial reform is being combined with stronger central government regulation of health care services (Sandberg & Sjöblom, 2022). In Sweden, such types of centralizing reforms have been pursued mainly by non-socialist governments, despite them having historically been in favour of subsidiarity, strong local self-government, and the adjustment of public service provision to local needs. Now new proposals include the nationalization of both schooling and the health system, which are currently local government responsibilities (Lidström, 2022). According to a recent report on local autonomy within the Nordic region, creeping centralization is taking place throughout the Nordic countries (Baldersheim et al., 2019).

2.4.1 Conclusions

The Nordic states are now facing a new situation regarding the formulation and implementation of public policy in which the established funding and direction for both their public sectors and their relatively generous welfare systems are being challenged. In the decades immediately after the Second World War there had been a period of extensive public financing of welfare through high levels of taxation throughout the Nordic region. This was followed by a liberalization and globalization phase where there instead was a new reliance on resources trickling down from the market economy. The Nordic countries are now experiencing the new challenges of the post-global era, where the consequences of climate change and increasing difficulties in international security are contributing to a much more uncertain situation.

Demands on resources continue to increase whereas revenues are uncertain. Both tax increases and reliance on free capitalism and international markets are seen to be risky strategies. There are some on the political left in the Nordic region who argue that the strong growth of private consumption during the last few decades may provide some additional scope for tax increases to finance joint commitments. Others, more to the right, claim that this would make the Nordic countries even more vulnerable to new disruptive forces and suggest instead that it is better to stay with liberalization policies, although their effects in terms of generating resources for public commitments may be smaller. It remains to be seen to what extent Nordic policymakers will choose one or the other of these paths to follow, or if an entirely new strategy will be developed.

We now proceed to the chapters within this volume that address specific public policy challenges in the Nordic countries. A number of crucial policy areas will be explored by experts in the field—looking in greater detail at the welfare state and political culture; globalization and liberalization; defence, security and foreign policy matters along with the impact of climate change and the environment. Together they provide an overview of some of the critical dimensions of contemporary Nordic public policymaking and explore the questions of "exceptionalism" and "inner diversity" found therein.

Acknowledgements Many thanks to Harald Baldersheim, Oslo University and Siv Sandberg, Åbo Akademi University, Turku for valuable comments on an earlier version of this essay.

References

Baldersheim, H., Houlberg, K., Lidström, A., Hlynsdottir, E.-M., & Kettunen, P. (2019). *Local autonomy in the Nordic countries: A report for the Norwegian association of local and regional authorities*. Universitetet i Agder/KS FoU.

Baldersheim, H., & Ståhlberg, K. (Eds.). (1994). *Towards the self-regulating municipality—Free communes and administrative modernization in Scandinavia*. Ashgate Publishing Limited.

Castles, F. G. (Ed.). (1993). *Families of nations: Patterns of public policy in western democracies*. Dartmouth Publishing Company.

Castles, F. G. (1998). *Comparative public policy*. Edward Elgar.

Esping-Andersen, G. (1990). *The three worlds of welfare capitalism*. Princeton University Press.

Hernes, H. (1987). *Welfare state and women power*. Universitetsforlaget.

Hesse, J. J. (Ed.). (1991). *Local government and urban affairs in international perspective*. Nomos Verlagsgesellschaft.

Hörnström, L. (2013). Strong regions within the unitary state: The Nordic experience of regionalization. *Regional & Federal Studies, 23*(4), 427–443.

Kautto, M., Fritzell, J., Hvinden, B., Kvist, J., & Uusitalo, H. (2001). *Nordic welfare states in the European context*. Psychology Press.

Kosonen, P. (2001). Globalization and the Nordic welfare states. In R. Sykes, B. Palier, & P. M. Prior (Eds.), *Globalization and European welfare states. Challenges and change*. Palgrave.

Lidström, A. (2007). Territorial governance in transition. *Regional and Federal Studies, 17*(4), 499–508.

Lidström, A. (2022). Den kommunala självstyrelsen—utmaningar och möjligheter. In Självstyrelse—*värd att värna*? Sveriges kommuner och regioner.

OECD. (2015). *In it together: Why less inequality benefits all*. OECD Publishing.

Palley, T. I. (2005). From Keynesianism to neoliberalism: Shifting paradigms in economics. In A. Saad-Filho, & D. Johnston (Eds.), *Neoliberalism: A critical reader*. Pluto Press.

Rose, R. (1984). *Understanding big government*. Sage.

Rothstein, B. (2000). Trust, social dilemmas and collective memories. *Journal of Theoretical Politics, 12*(4), 477–501.

Sandberg, S., & Sjöblom, S. (2022). Den kommunala självstyrelsen i Norden. In S. Godenhjelm, E. Mäkinen, & M. Niemivuo (Eds.), *Förvaltning och rättsäkerhet i Norden*. Svenska litteratursällskapet i Finland. Appell förlag.

Sellers, J. M., Lidström, A., & Bae, Y. (2020). *Multilevel democracy. How local institutions and civil society shape the modern state*. Cambridge University Press.

Sipilä, J., Anttonen, A., & Kröger, T. (2009). A Nordic welfare state in post-industrial society. In J. L. Powell, & J. Hendricks (Eds.), *The welfare state in post-industrial society*. Springer.

Tanzi, V. (2011). *Government versus markets: The changing economic role of the state*. Cambridge University Press.

Tanzi, V., & Schuknecht, L. (2000). *Public spending in the 20th century: A global perspective*. Cambridge University Press.

Torfing, J., Lidström, A., & Røiseland, A. (2015). The Scandinavian regional model: Accounting for the shift from convergence to divergence. *Scandinavian Journal of Public Administration, 19*(4), 7–28.

Anders Lidström is Professor Emeritus of Political Science at Umeå University, Sweden. He specializes in local, regional and urban democracy and governance. Anders Lidström was the convenor of the ECPR standing group on local government and politics from 2010–2016; Head of the Department of Political Science, Umeå University from 2010–2016 and Chair of the Swedish Political Science Association from 2015–2017. He is currently the Swedish member of the Group of Independent Experts on the European Charter of Local Self-Government, Council of Europe. Recent publications include *Multilevel Democracy: How Local Institutions and Civil Society Shape the Modern State* (with Jefferey Sellers and Yooil Bae), Cambridge University Press, 2020; a special issue of *Journal of Urban Affairs* on "The Citizens in City-Regions" (No 1, 2018, edited with Linze Schaap) and, Sverigedemokraternas regionala politiska kultur i Skåne och Blekinge. Vad ligger bakom partiets exceptionella regionala framgångar? (Umeå Working Paper in Political Science, 2025:1–with a summary in English).

Part II
The Contributions of Political Culture and the Welfare State

Chapter 3
Nordic Political Culture: Democracy, Emancipative Values and the Rise of Populism

Bi Puranen and Lennart Joe Brunkert

Abstract The Nordic countries have long been considered beacons of liberal democracy, marked by strong social support structures, high institutional trust, and deeply ingrained emancipative values. However, the rise of populist parties with authoritarian tendencies disputes the conventional view of Nordic exceptionalism. This chapter explores the paradox of the Nordic political landscape, where high levels of support for liberal democratic principles coexist with a growing inclination toward populist and nationalist parties. The findings suggest that the Nordics are truly exceptional in their cultural orientation towards emancipation and liberal democracy. Apart from Finland, this exceptionality is also reflected in political profiles of their populist parties. Nordic populism focuses less on the reestablishment of traditional family models, is less discriminatory towards queer people and features weaker anti-elitism than their central European counterparts. Despite these differences the growing power of Nordic populism and its strong anti-immigration stance are a worrisome development which contradicts the liberal mindset of the Nordic cultures and needs to be monitored closely.

Keywords Values · Democracy · Emancipative values · Populism · Nordic states · Exceptionalism · World values survey

3.1 Introduction

The idea that the Nordic countries—often regarded as crusaders of democracy, equality, and social welfare—could also harbor relatively high levels of authoritarian values may seem paradoxical. However, this is what seems increasingly evident in this region of the world. In recent years, there has been a rise in populist parties that have

---

B. Puranen (✉)
Institute for Future studies, Stockholm, Sweden
e-mail: bi.puranen@worldvaluessurvey.org

L. J. Brunkert
Leuphana University, Lüneberg, Germany

D. C. Nord (ed.), *Contemporary Nordic Public Policies*, Springer Polar Sciences,
https://doi.org/10.1007/978-3-032-11292-7_3

espoused racist and anti-immigrant views, which suggest that the Nordic societies are not as immune to these tendencies as has been traditionally suggested. Determining the relative balance between liberal democratic values and certain authoritarian beliefs seems to be central to understanding what is shaping Nordic political culture today.

Accordingly, we intend to analyze whether traditionally strong democracies such as the Nordics may be better equipped to handle the rise of populist parties and to reconcile certain value contradictions within their communities. To better understand democratic developments, we look at individuals' notions of democracy and how much these are deemed essential in society. Liberal notions of democracy include civic freedoms where people can make their own choices within the limits of the laws that are democratically decided. These laws protect them and give them a voice and a vote in politics. They also protect minorities and gender equality.

Contrary to this, are certain authoritarian notions of democracy that suggest that citizens need to obey their rulers and that the military and/or church should be able to influence the government. Clearly, these latter ideas do not resonate well with most established views of democracy. Nevertheless, there is a surprisingly large number of European citizens who are unable to clearly delineate democratic from authoritarian practices, as we will see.

Traditionally, the Nordic countries are at the forefront of liberal democratic values and score high according to all traditional notions of liberal democracy. How, then, can we explain the growing number of voters in the Nordic countries who now support populist parties with authoritative agendas?

One way of finding out is to ask people in representative surveys about their perceptions of what constitutes a democracy and how they relate to questions of equality, freedom, and self-determination. This is exactly what the World Values Survey (WVS) has done since 1981 by conducting face-to-face interviews (For more information, see www.worldvaluessurvey.org). Since its launch, this large-scale NGO survey, coordinated by researchers from all participating countries, has collected data from over 600,000 respondents, reflecting the thoughts, values, and ideas from societies representing approximately 90% of the world's population. In the most recent wave (Wave 7, completed in 2022), around 150,000 individuals were interviewed.

The well-known WVS Culture Map (see Fig. 3.1) indicates that the Nordic countries are outliers, being highly secular and demanding of individual freedom expressed in terms of emancipative values. As such, they can be seen as "exceptional" when compared with other societies. Political culture research suggests that these value-dimensions—being emancipative and secular—form the baseline for enduring democratic systems and a liberal–democratic mindset within these Nordic societies.

However, despite being on the emancipative and secular end of the spectrum, a surprisingly large share of the Nordic population has, over the past few decades, voted for various populist parties. Although these parties have not received a majority of the vote, their political support has, nonetheless, hovered in some instances around 15–20% of the Scandinavian electorate (SCB statistics; Jungar & Jupskåps, 2014).

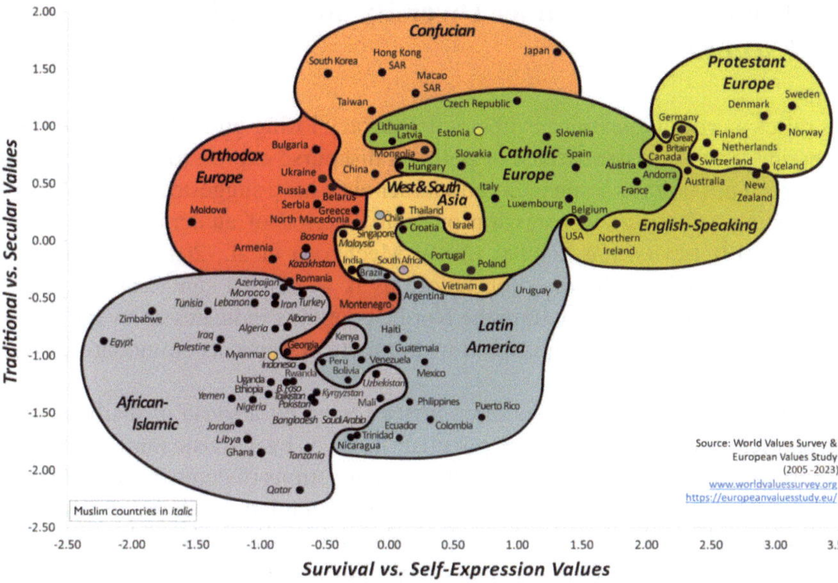

Fig. 3.1 The world values survey culture map, 2023

Over time, the balance of political power in the region has shifted. In Denmark, public dissatisfaction is now channeled through three different populist political parties. In Sweden, the government is now ruling with support from the populist Sweden Democrats. In Iceland, the recent coalition government resigned because of internal strife between the various parties who failed to secure the trust of the public (government.is, 2024).

The former Danish Social Democratic leader, Jens Otto Krag, has remarked that the success of these populist parties likely represents a form of voter retribution. He argues that voters feel confused and disappointed by the established politicians. This “revenge of the voters” has created an opening for populist parties to relay the electorate’s discontent (Lidegaard, 2011; Lindroth, 2016, 2024). It has also been argued that the sometimes-blurred borders between liberal and authoritarian notions of democracy are reinforced by the media, and not least by the rise of social media, which can lead to a normalization of populist parties and their narratives (Norris & Inglehart, 2019).

3.2 The Concepts of Inner Diversity and Nordic Exceptionalism

This raises a crucial question: Is liberal democracy losing ground, or are we simply witnessing ripples on the surface of political discourse expressed in public discontent? Or is it possible, as some international scholars have suggested, that we are truly experiencing a democratic recession or even a period of autocratization? (Foa & Mounk, 2017; Levitsky & Ziblatt, 2018; Mechkova et al., 2017; Mounk, 2018; Schäfer et al., 2023). The question before us is whether a significant increase in the vote for populist parties within the Nordic countries indicates an overall decline in democracy or simply an adjustment within party systems accommodating diverse political voices.

Populism within democracies often arises fueled by a sense of disillusionment with political elites. However, when populism is driven by nativism (anti-immigration and nationalism) and also fueled by authoritarian notions of democracy, it may challenge established democratic norms. This may potentially undermine democratic institutions (Alexander & Welzel, 2017; Appelbaum, 2020; Brunkert, 2022; Brunkert et al. 2024; Carter, 2018; Czarnek et al., 2019; Gidron & Hall, 2020; Goodhart, 2017; Haerpfer, 2018; Norris, 2011, 2012; Norris & Inglehart, 2019). This may be especially pertinent in the Nordic context, where populist parties have grown significantly strong within traditionally stable democracies (Jungar & Jupskås, 2014; Mudde, 2017, 2019; Puranen, 2022; Rydgren, 2018; Wikforss, 2021).

The question here to be examined is whether current Nordic populism is also "exceptional." We hypothesize that Nordic populism might be a localized version of populism seen elsewhere in Europe but being less anti-EU, less misogynist and less traditional in character. Nevertheless, we expect it to feature other key elements of European populism, including anti-elitism and people centrism (Crum & Oleart, 2023; Fassin, 2017; Kaltwasser & Mudde, 2017; Mudde, 2004). Nordic populism also allegedly seeks to protect the strong welfare systems of the region and the homogeneity of small nations. In other words, it is a populism with a certain nationalist image but without radicalism. It does contain, however, a strong anti-immigration agenda.

3.3 Towards More Inner Diversity—Towards More Populism?

The concept of inner diversity may be of use in better understanding the anti-pluralist populism that seems to be emerging in the Nordic region. In this case, we are not referring to the inner diversity in public policies among the Nordics countries that is discussed in other chapters of this volume. Rather we are considering the nature of internal diversity found within different societies of the world. Some societies can harbor a variety of differences and coexist with several cultural, ethnic, religious, linguistic, and social groups within their population. Others are in constant fights both within the country and with neighbors. The concept of inner diversity as utilized here suggests that a range of identities, traditions, and worldviews contribute to a nation's societal structure. It highlights the multifaceted nature of most countries and emphasizes the importance of inclusivity, mutual respect, and unity among diverse groups. "Inner diversity" is often considered a strength in many societies as it can foster innovation, cultural richness, and resilience.

The Nordic countries, however, developed as relatively small, homogenous populations with shared cultural, social, and institutional values. Characteristic of these societies was the degree of trust existing since the Viking Age and their (relatively) inclusive decision-making practices. Over the centuries, common social structures were built. Later, social policies based on equality, education, and welfare further strengthened this sense of trust and community. (Barrling, 2024; Harrison & Huntington, 2000; Rothstein, 2005, SOU, 2020; Tolinson-Ting, 2020). Over recent decades, increased globalization, migration, and cultural changes have introduced new layers of inner diversity within the Nordic societies and challenged centuries-old national narratives. These forces have brought with them greater ethnic, cultural, and ideological differences within the population. While many Nordic citizens embrace this diversity, others perceive it as a challenge to traditional norms or values. Support for populist parties can be seen, in part, as a reaction by some groups to this growing diversity. These parties often capitalize on concerns about cultural preservation, economic competition and political influence. They often frame such inner diversity as a potential threat to the cohesion and trust that Nordic societies prize.

Despite the growing populist vote, the Nordic countries remain internally strong among European countries with the highest levels of inter-individual trust and trust in institutions. Yet, high levels of general trust in a society does not imply the absence of tension or disagreement. Instead, it provides a baseline willingness to cooperate, even among individuals with different backgrounds. The coexistence of trust and populist voting reflects the complexity of internal diversity in Nordic countries today. Even those citizens who vote for populist parties may still trust institutions like the judiciary or healthcare systems. Their vote may not signify a complete rejection of the system but rather a demand for addressing perceived issues within it—such as migration policies, economic inequality, or political representation.

Robert Putnam's' studies on the impact of migration and diversity provide a framework for understanding how inner diversity can create challenges in the Nordic countries and elsewhere (Dalton & Welzel, 2014; Lindroth, 2021, 2022; Putnam, 2007; Radkiewicz, 2017). The values and social norms of the newly arrived migrants from non-Western societies creates high stress not only on the migrants themselves but also on the whole of society (Puranen, 2019, 2020, 2021, 2023). The existence of inner diversity challenges societies to adapt their trust frameworks to include these new groups. In the Nordic context, this involves negotiating between the desire to preserve traditional societal models and embracing the new opportunities that diversity offers. Support for populist parties is part of this negotiation process.

Still, we would expect the Nordic societies, overall, to be less xenophobic and to behave more inclusive fashion than other communities given their above-average support for the protection of civil liberties, their focus on emancipative values and their trust in other people. And indeed, studies of the "Nordic model" emphasize the balance that exits between social equality, democracy, and economic openness, even under the strain of increased inner diversity (Kuhnle & Alestalo, 2017; Kuhnle & Olsson, 2004; Kvist & Greve, 2011, Kvist et al., 2012; Mudde, 2017). One representative example of this openness, trust and respect for individual liberty lies in the Nordic countries' strong focus on emancipative values. This we will discuss in the final section of this chapter.

3.4 Authoritarian Versus Liberal Notions of Democracy

As a first step to better understand the differences between liberal and authoritarian notions of democracy, we will make use of the World Values Surveys data from its 7th Wave. This allows us to compare results from the Nordic region with other findings from across the globe. The comparative method is a well-established approach in the social sciences science that helps to explain why certain societal features are dominant in one case and not in another (Della Porta, 2008). In addition, we rely on the Varieties of Parties Project data, which uses expert coding to evaluate the characteristics of countries' parties, including their degree of populism (Lindberg et al., 2022). As we are focusing on Nordic exceptionalism, we compare the Nordic countries with other societies from Europe and beyond. We do this to better understand the historical and contemporary influences that distinguish Nordic political culture from the rest of Europe and the world.

To measure liberal versus authoritarian notions of democracy, we follow the example of Kirsch and Welzel (2019) and utilize three indicators from the what *democracy means* portion of the WVS (www.worldvaluessurvey.org) survey.[1] We would expect that citizens of established democratic countries should be able to clearly identify free and fair elections, civil liberties and the principle of equality as prime features of liberal democracy. At the same time, they would reject army rule, religious laws and unconditional obedience to authorities as undemocratic features. Figure 3.2 summarizes the average relevant score for a number of countries. These scores are, in turn, grouped into separate cultural zones. The differences between these zones are striking and will be discussed below.

Before considering these, however, we need to point to the difference in the scale-ranges for the two notions of democracy. With regard to the authoritarian view of

[1] The liberal notions of democracy consist of the items V249: *Women have the same rights a* men; V246: Civil Rights protect people's liberty against oppression and V243: *People chose their leaders in free elections.* Authoritarian notions of democracy consist of V248: *People obey their rulers*; V245: *the army takes over when government is incompetent* and V242: *Religious authorities interpret the laws.* The items are each scaled from 1 to 10 with 10 meaning that the statement is an essential feature of democracy. We created the index by averaging the values across the three components for each notion of democracy. For an extensive discussion of these indices, please see Kirsch and Welzel (2019).

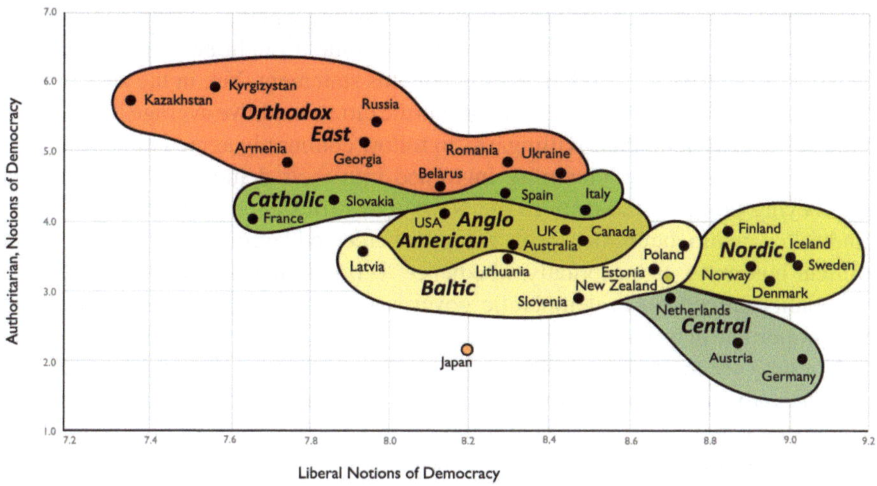

Fig. 3.2 Authoritarian versus liberal democracies—Nordic, Baltic, Catholic, Central EU, Anglo-American and the Orthodox East

democracy, we find average values among the countries sampled ranging from a strong rejection of these values (around 2.0 for Germany) to strong support for them (around 6 for Kirgizstan). For liberal notions of democracy, on the other hand, we only find average values above 7.0 for all the countries sampled. Accordingly, we can conclude that most societies around the world have a relatively good knowledge of what liberal democracy traditionally means. In the central European and Nordic countries there is almost universal, absolute agreement that elections, civil liberties and gender equality are central pillars for the concept of democracy. The main differentiator among the countries, seems to be on how strongly citizens also include elements of an authoritarian concept into their individual view of democracy.

3.5 The Nordic Countries

Scores for liberal democratic values are consistently high across all Nordic countries, ranging from 8.83 to 9.14. Iceland and Sweden show particularly strong support for these ideas. These scores are the highest in the world and reflect a deeply ingrained commitment to pluralism, freedom, and democracy. The Nordics are also known for having strong social safety nets, well-functioning political institutions, and active civil societies. All of which are consistent with liberal democratic values.

However, it may seem contradictory then that the authoritarian scores in the Nordic countries are generally higher than in Germany or Austria, especially considering the strong democratic traditions of the Nordic countries. The reason for this lies in one of the three components of the index (see SOM, A1). Citizens of the Nordic countries consistently score high on the variable question asking whether obeying one's rulers is an essential characteristic of democracy. For some it might be surprising that all the Nordic countries score relatively high on this statement. Yet, in the case of these societies, this obedience results from the Nordic citizens' above average level of trust in the political system, their willingness to follow commonly agreed upon rules, and the high degrees of accountability ingrained into their democracies. This means that obeying political authority in the Nordic countries is not considered an act of blind followership, but rather stems from the perceived legitimacy of their systems' political outcomes. Conceptually, obedience to rulers may seem to contradict the impartiality embedded in democratic governance, where procedures and laws are more important than individuals. However, if some survey respondents (including the Nordics) see their elected officials as highly trustworthy, they may extend their willingness to obey to these individuals as well. Our sampling of authoritarian notions of democracy may have inadvertently capture this as support for authoritarian ideals. The reality, however, is that it is more an indication of an individual's willingness to perceive a public office, and the current incumbent, as legitimate outcomes of the political process and thus worthy to be obeyed.

The Nordic countries have historically had a strong sense of social cohesion and societally agreed upon rules, which lend legitimacy to their institutions. They score the highest among those sampled on trust, measured by an index consisting of generalized trust and trust in people you meet for the first time. Among the 64,607 worldwide respondents to the survey, some 22% trust their fellow citizens. In the Nordic countries, however, agreement with this statement rises to an astonishing 55%, highlighting the high levels of societal cohesion within the region. Likewise, their degree of confidence in political institutions is above average. Their confidence in the parliament, in the government, the courts, the armed forces, the churches, the universities, the police and the civil services is among the highest found in this study. While this trust is often directed toward democratic institutions, it can also be regarded as a broader embrace of institutional authority in general, including state control and regulations. This Nordic acceptance of state intervention might obscure the line between liberal and authoritarian values—especially when people see strong state leadership or regulatory power as positive in maintaining welfare and societal equality. High trust in the government's ability to make decisions in the best interest of the community can lead some people to accept government-imposed restrictions or regulations without perceiving them as a threat to democracy.

So, paradoxically, the slightly higher authoritarian scores among the Nordic countries when compared with those found in Germany or Austria, could be attributed to a combination of cultural factors including trust in institutions and a consensus-oriented political systems. All of these societies tend to be more comfortable with state authority and centralized governance due to their strong welfare-state models and historical reliance on their governments to maintain social cohesion (SOU, 2020). Especially in the case of the Nordics, societal homogeneity, common good and consensus orientation increase the willingness to obey authoritative decisions.

3.6 Introducing Emancipative Values into the Analysis

Already in this chapter, we have made mention of "emancipative values", which are closely related to the WVS cultural maps' secular and self-expression values. Emancipative values (EV) measure attitudes toward equality, autonomy, choice, and voice among the population and have been found to encourage demands for liberal democracy, mutual respect, and individualism (Welzel, 2013). Accordingly, we expect that—from a comparative perspective—high levels of emancipative values would encourage liberal notions of democracy. In turn, these notions help to create and maintain a democratic system that institutionalizes the civil liberties that emancipative values demand.

The Emancipative Values Index consists of four sub-indices tapping respondents' attitudes to autonomy, equality, choice, and voice:

Autonomy includes questions on what the respondent prioritizes when bringing up a child. Will the objective be to develop your child into an independent and creative being, or do you want to prioritize discipline and obedience with respect to the child?

Equality asks the respondents to evaluate if men should have priority in jobs, university/education, and political leadership.

Choice is about the individual's right to decide for him/herself on issues such as marriage, divorce, abortion and sexual preferences.

Voice, relates to whether one's voice heard. Can it exert influence in one's workplace, local community or with government.

These sub-indices create the Emancipative Index, which is plotted together in Fig. 3.3 with our index of liberal—versus—authoritarian notions of democracy.

We suggest that "emancipative values" are most influential, exerting a two-fold 'enlightening' effect in (a) making people recognize the contradiction between liberal and authoritarian notions of democracy and (b) turning them against authoritarian notions. In a nutshell, the prospects of democracy are bleak where emancipative values remain weak (Kirsch & Welzel, 2019; Welzel, 2013).

When plotting liberal/authoritarian values against the level of emancipative values from sampled countries, it becomes apparent that the level of EV is decisive in determining the level of liberal democracy and its stability. Both value dimensions go hand-in-hand, because growth in one dimension benefits growth in the other. Individuals who value individual freedom, such as the liberty to love and live together with whomever they please, will most likely advocate a political system that enshrines these civil liberties into its constitution. They will certainly not like to have church laws interfere in their personal life. Similarly, individuals who do not see obedience to authority as an important factor in their political system would be more likely to also educate their children in ways that promote autonomy over dependence.

The Nordic countries are at the very top in the contemporary world with regard to these emancipative values (See Fig. 3.4). Independence, trust, respect for privacy and non-interference in other peoples' lives rank very high within the political cultures of

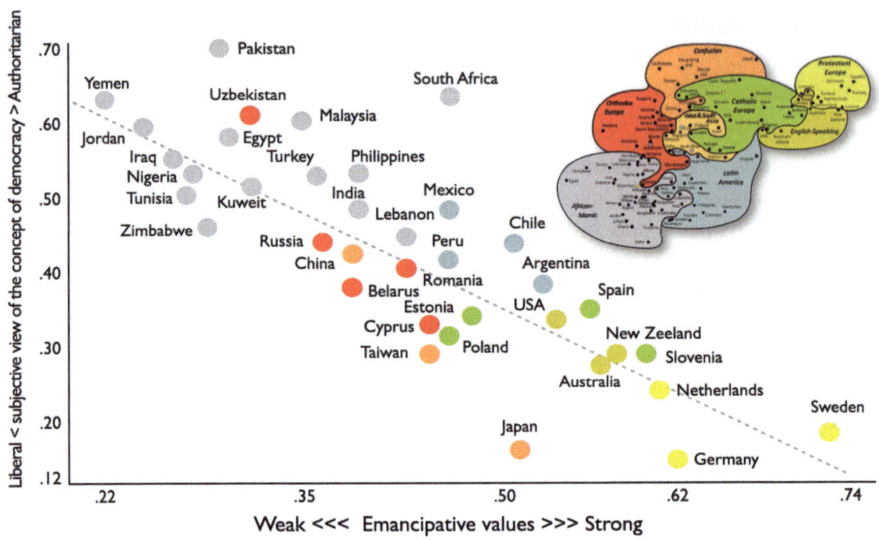

Fig. 3.3 The association between the country's individual liberal and authoritarian notions and their emancipative values

these societies. This can also be seen in their comparatively late marriage ages, lower birth rates among women, and a high levels of female employment. To empirically validate this point we present data from 39,997 interviews in all represented countries and show their respective shares of individuals with *low, intermediate*, and *high* emancipative values.

The disparity in emancipative values (EV) between the Nordic and Western European countries and of those nations like Russia, Kazakhstan, Uzbekistan, Tajikistan, Kyrgyzstan, can arise from a combination of many factors. The latter likely share a legacy of authoritarianism rooted in the Soviet era, where individual freedoms were subordinated to collective goals and state control. This inheritance has

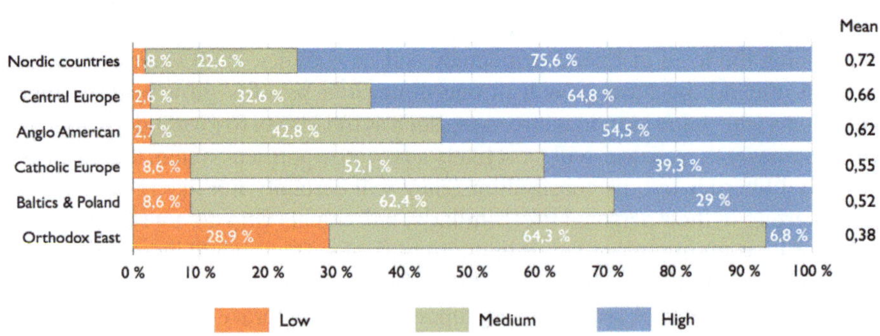

Fig. 3.4 An analysis of the emancipative values index within cultural zones

stifled the development of emancipative values such as self-expression, individual autonomy, and participatory decision-making. Today, many Central Asian countries and Russia are governed by regimes that suppress political opposition, limit civil liberties, and control media narratives. These environments discourage the development of emancipative values by restricting freedom of thought, speech, and political participation.

Additionally, emancipative values tend to grow with economic security. When basic needs are unmet, individuals frequently prioritize survival over self-expression and the protection of civil liberties becomes less important. We find the same result, where education levels are uneven in societies and when limited emphasis is placed on critical thinking and civic education. Restricted access to independent media seems also to reinforce authoritarian narratives and hinders the dissemination of emancipative ideals. Overcoming this multitude of disparities found between these countries and those of the EU and North America will require profound political and economic reforms as well as substantial shifts in cultural attitudes and educational systems (Gessen, 2017; Snyder, 2018). By comparison the Nordic countries already have a head-start. Their consensus-based style of decision-making, their valuation of female independence, and their Protestant faith have embedded an emancipative mindset within these societies and propelled them toward establishing democratic institutions.

3.7 Populist Parties in the Nordic Countries and Beyond

In the first part of this chapter, we compared the Nordic countries views of liberal versus authoritarian notions of democracy with a wide range of different societies. These value profiles help to make sense of underlying political preferences among the specific national populations. However, they do not tell us much about the actual operation of political systems in these countries. Hence, we should ask whether more authoritarian and less liberal notions of democracy also allow for the creations of authoritarian parties and if there is a specific Nordic form of contemporary political populism. In this section, we will compare the main populist parties found in the Nordic countries. These are the Danske Folkeparti (Danish People's Party), the Sverigedemokraterna (Sweden Democrats), the Perussuomalaiset (True Finns) and the Fremskrittspartiet (Progress Party, Norway).

To identify the degree of populism evident in these parties, we use the last available coding from the V-Party dataset (referring to elections between 2016 and 2019) which uses expert ratings to identify a set of party characteristics (Lindberg et al., 2022). We are most interested in those characteristics that are traditionally associated with populist parties. As mentioned above, these parties often stress elements of people-centrism and anti-elitism. They also assume a unidimensional will of the people and advocate a majoritarian understanding of democracy which does not take minorities and foreigners into consideration. Additionally, such parties establish an *us-vs-them* narrative where the established political elites are portrayed as greedy, corrupt and power-hungry. From their perspective, the purity of the people stands against all of this (Mudde, 2004).

In addition to these main features of populist parties, there also exist a variety of ideological orientations held by members. The V-Party dataset includes items measuring the degree of anti-immigration attitudes, anti-LGBTQI + feelings, perceptions of cultural superiority, and a belief in a traditional family-pattern with stay-at-home mothers. We use these indicators to compare populist parties within the Nordic countries with those of other European countries where populism has already taken a significant foothold. In line with their existing emancipative and liberal democratic values, we would expect the countries of Sweden, Norway, Finland and Denmark to feature a different style of populism than for example seen in Poland and Hungary. There, populist leaders have actually governed and have been backed by less liberal societal values.

And indeed, we can clearly see such differences when we compare the Nordic cases with other European examples of populism. (See Fig. 3.5). The anti-LGBT rhetoric used by populist parties in Denmark, Norway and Sweden is less pronounced than is seen in other European countries. Only the True Finns come close to the German AfD party in their hate for sexual liberty. Likewise, in the three Nordic countries, we find that their populist parties are far less traditional in their view on gender roles. Again, however, the populists of Finland tend to follow a different path.

Anti-Elitism (1) is the strongest value in Germany. People Centrism (2) is the most common response in Sweden, Finland and Poland. Anti-Immigration (3) scores the highest in Norway and Cultural Superiority (5) is the dominant answer in France.

Looking at the remaining European countries in Fig. 3.5, we can clearly see that populism in these societies is not only about opposing immigration and claiming

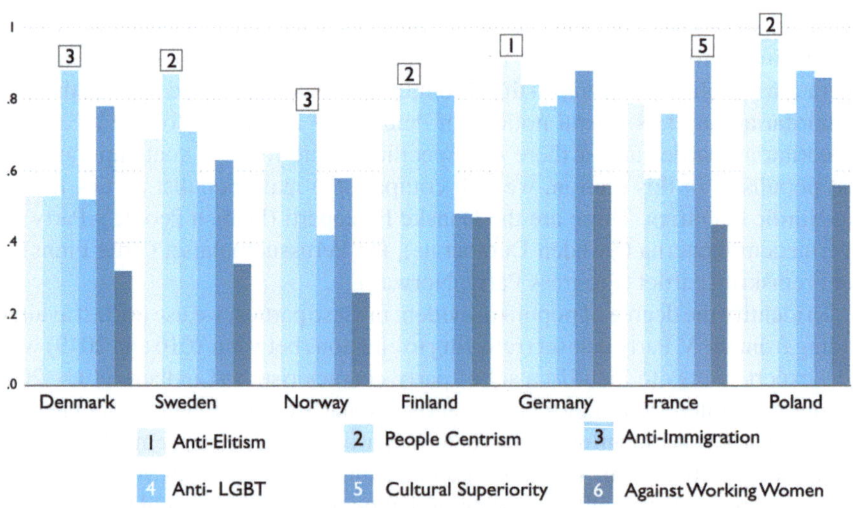

Fig. 3.5 Main populist parties in the Nordics and some European countries

to “take back power” for the people. Populism in these countries seems also to be about the exclusion of LGBTQ + people from society and the return of women and children to a “traditional” family arrangement.

With regard to other dimensions of populist belief, there seems to be a bit more diversity in the attitudes of populists within each Nordic country. In Denmark, the defining positions of populist there seems to be a rejection of immigrants along with a strong nationalist view of the cultural superiority of Danes. There is less of a focus on anti-elitism or people-centrism. The “Swedish Democrats,” on the other hand, rely most strongly on a people-centered rhetoric in their appeals and a commitment to “recapturing the state for the people.” Norwegian populism, like its Danish counterpart, is also immigration centered. However, its other features are much less pronounced. The populist True Finns again clearly stand apart from their other Nordic counterparts. Their attitudes are much closer to the Central European variants of populism. Yet there is a less pronounced narrative on cultural superiority or the downplaying of women in the workforce.

Iceland has no history of populist far-right parties so far. According to the *Authoritarian Populism Index* (2024), support for populist parties in Iceland is among the weakest in Europe. In the 1990s, Iceland’s number of foreign-born residents was very low, under 3% of the population. However, following the EU’s subsequent expansion, immigration—primarily from Eastern Europe— has increased significantly. Today, more than 18% of Iceland’s population is foreign-born, one of the highest shares in the Nordic region (Statistics Iceland, 2024).This development has prompted the Progressive Party (Framsóknarflokkurinn), one of Iceland’s established parties, to make anti-immigration policy statements in recent years. Iceland, however, has not witnessed the emergence of new strong nationalist or anti-immigration parties as seen in the other Nordic countries. Nonetheless, discussions continue within the established political arena with respect to how to better integrate immigrants into Icelandic society, particularly with regard to language learning and citizen participation.

In summary, the strong anti-pluralist populism found in many other European countries does not seem to fit to the overall Nordic value profile. Instead, it seems that a “milder” form of populism has emerged in the Nordic countries when compared to the rest of Europe. It is exclusionary towards foreigners but less opposed to societal inner diversity in terms of sexuality and gender. This does not make it less dangerous, but shows that (a) prevailing emancipative values can also influence those actors most opposed to them and that (b) liberal democratic values have deeply engrained the protection of civil liberties into the shared values of most Nordic populations—at least for those belonging to the perceived *in-group*. Interestingly, the success of the populist True Finns seems to deviate from both of these observations. This makes our findings not generalizable for all Nordic countries.[2] This also suggests that the degree of Nordic political cultural exceptionalism is not uniform throughout the region.

---

[2] Turning to additional countries for comparison does not change our findings, since the Anglo-Saxon bi-party systems (UK, USA) lack the diversity to identify an outlier populist party. Russia and other Eastern countries lack all competition between parties and Latin American parties have traditionally been more populist, though often with a leftist outlook.

3.8 Discussion

The influx of migrants and refugees to Europe, particularly during the 2015 refugee crisis, has significantly influenced the rise of right-wing populists throughout the continent. Immigration is often framed as a threat to national identity, culture, and security by populist leaders. They also advocate stricter border controls and policies that prioritize native populations. European Populism also thrives on the perception that political elites are corrupt, out of touch, and serve the interests of global corporations and international organizations rather than ordinary people.

Many reasons have been suggested as fundamental causes of this expressed skepticism toward liberalism and democratic institutions. One of the leading theories suggests that economic inequality and discontent, exacerbated by factors such as globalization, automation, and austerity policies, have left many people, particularly in rural areas, and among the working class, feeling left behind. These economic challenges have created fertile ground for illiberal movements, which often blame elites and international institutions for the population's economic difficulties.

Nevertheless, Europe remains a continent of liberal democracies. However, they may have to compromise to keep their national populist opponents in check. The mixed dynamic of mostly liberal—and partially authoritarian values—within these contemporary societies reflects a broader European struggle over the future of democracy. This is particularly the case in those regions facing economic challenges, political disillusionment, or historical legacies of authoritarianism.

While most citizens generally express support for democracy, the WVS' data spanning 40 years of interviews shows that their understanding of what democracy means can vary widely. A vast majority of the respondents value the classical idea of democracy (fair elections, representation, and freedom) and support the institutions and norms that are essential to a functioning liberal democracy. For many European countries, liberal democratic values retain their strength against the backdrop of rising populism and illiberal movements. The populations in many countries express a clear preference for inclusive, rights-based democratic governance rather than centralized, authoritarian rule. This suggests the resilience of liberal democracies when they have the tools and will to adapt and respond to challenges confronting them. They seek to maintain their core principles of freedom, equality, and justice.

Overall, authoritarian scores in different European countries from Wave 7 of the WVS survey remain comparably low. This suggests that while populist and authoritarian movements may have some political influence, they have not managed to supplant the core democratic values in most societies. Still, there are some illiberal governments in Europe that have successfully undermined certain core democratic norms including independent judiciaries, civil societies, and a free media. Sometimes they have been able to do this without facing widespread domestic resistance (Appelbaum, 2020).

Our results suggest that the Nordic countries are slightly more supportive of certain authoritarian precepts of democracy than some of their central European partners. While this may be a surprising result, we can trace the Nordics' response to their sense of *well-justified obedience.* It can be argued that in their case, the Nordics' obedience is voluntary, conditional, and rooted in the legitimacy of the democratic process. Nordic democracy allows for political authority to be checked, questioned, and eventually replaced through lawful means, providing a balance between order and freedom. The key distinction to be focused on in the Nordic case is that in their democracies, obedience is part of a broader social contract, where citizens maintain the power to influence and change the laws that they are asked to follow. Obedience in such a democracy is conditional, grounded in the principles of participation, accountability, and the rule of law. Obeying the law in a Nordic democracy does not entail blind obedience to authority but rather an acknowledgment that laws are meant to protect the rights and well-being of all citizens.

The Nordic countries' exceptionally high scores on both emancipative and liberal democratic values may well stem from their unique wartime experiences and subsequent social development. While Denmark and Norway faced occupation, their post-war reconstruction emphasized social cohesion and egalitarianism. Finland, navigating a precarious position between the West and the Soviet Union, developed a pragmatic approach to governance that balanced external pressures with the need for internal stability. These factors may have contributed to a nuanced blend of liberal and authoritarian elements in Nordic democratic values. Additionally, the degree of emancipation—a society's emphasis on individual freedom and autonomy, self-expression, and participatory decision-making—helps to explain these differences. Societies with higher levels of emancipation tend to prioritize liberal democratic values, while those with moderate levels may integrate elements of authority into their conception of democracy.

Even before World War II, the Nordic countries began pioneering welfare state policies, emphasizing equality, education, and social security. This created a shared ethos of collective responsibility and egalitarianism. Despite Sweden's neutrality, it aligned culturally and politically with its Nordic neighbors during the reconstruction period after the war. The rise of the United Nations and the Nordic Council reinforced regional cooperation and shared values. Being relatively small in terms of population, the Nordic countries have historically relied on multilateralism and diplomacy, fostering open, democratic societies that prioritize cooperation, human rights, and rule of law.

High levels of general trust allow the Nordic countries to manage their societal inner diversity pragmatically. While people might view certain changes in their communities skeptically, they still trust in their institutions to mediate conflicts, uphold fairness, and implement reforms. For example, trust in the Nordic justice systems or welfare institutions may remain high even among those critical of immigration policies or global influences. The Nordic Model's resilience lies in its ability to absorb differences without eroding trust. Institutions actively work to integrate diverse groups, and social norms encourage dialogue and compromise. This Nordic trust creates opportunity for a broad spectrum of views, including populist critiques, to coexist without undermining the overall social fabric.

3.9 Conclusion

In conclusion, the interaction between liberal and authoritarian values, shaped by historical experiences and levels of societal emancipation, demonstrates the complexity of democratic ideals across cultures and their differing impacts. The legacy of significant historical events, like World War II, and the unique trajectories of different regions of the world continue to inform their democratic frameworks and values today. Modern crises challenge established political systems which have for most of their existence been dominated by relatively homogenous populations.

High levels of general trust in Nordic countries do not negate the presence of inner diversity within their societies but rather shape how that diversity is navigated. Trust acts as a stabilizing force, enabling these societies to adapt to demographic and cultural changes while maintaining social cohesion. Contemporary Nordic populism, in this context, is less about rejecting trust and more about negotiating its boundaries in a rapidly changing landscape. Nonetheless, some Nordic populist narratives often oppose pluralist visions of both present and future societies in the region and need to be constantly checked by the remaining political parties and the broader electorate if the notion of inner diversity within Nordic societies is to be maintained.

The rise of right-wing populism, even in strong democracies like those of the Nordic region, forces us to reconsider simplistic narratives about democracy's decline. A close examination, suggest that Nordic democracy faces neither a straightforward regression nor just a series of minor disruption. Instead, it presents a portrait of an intricate and evolving system, which must balance competing interests of inclusivity, representation, and stability in society. Whether these developments are a true sign of democracy losing ground or merely adjusting to emerging challenges is a question that will continue to drive debate among political thinkers. We share the conclusion earlier cited from former Danish prime minister Jens Otto Krag who suggested that the success of populist parties represents a form of voter retribution against the established order. However, this type of retribution needs to be carefully studied and not allowed to exceed certain thresholds. Because as soon as populists start to govern and become more than reservoirs of the dissatisfied, they often start to attack the societal inner diversity of their countries and begin defining who belongs and who does not. In suggesting a return to "Bullerbyn" they are calling for a reduction of civil rights, the deportation of new citizens and a return to cultural practices that discriminate at least half of the population (women), if not more (LGBTQI +).[3] Such attitudes and actions may not be readily accepted within most contemporary Nordic communities.

When politics and politicians fail to address pressing problems, voters, for legitimate reasons, express their democratic rights to object. Populist parties can channel

[3] The term "Bullerby" in this context refers to an idealized, nostalgic, and overly simplistic vision of a harmonious, homogeneous society. The term is derived from the fictional village of Bullerbyn, created by the Swedish author Astrid Lindgren in her children's books. In these stories, Bullerbyn is portrayed as a small, idyllic rural community where life is simple, everyone knows each other, and harmony reigns.

the electorate's discontent within the democratic system even if it means giving the established political actors some tough criticism. This freedom of speech is a virtue of true democracy and is prevalent throughout the Nordic region. However, such opposition must not fall victim to intolerance. Democracy is a system where decision-making is based on rational deliberation, public debate, and dialogue. Emancipative values have shown to be the strongest vaccine against excessive populism. Where emancipative values are strong, democracy will remain strong. This is the dominant relationship that should continue within the Nordic countries.

References

Alexander, A. C., & Welzel, C. (2017). *The myth of deconsolidation: Rising liberalism and the populist reaction.* ILE Working Paper Series.

Appelbaum, A. (2020). *Twilight of democracy: The seductive lure of authoritarianism.* Doubleday.

Authoritarian Populism Index, 2024. https://www.epicenternetwork.eu/publications/timbro-authoritarian-populism-index-2024/

Barrling, K. (2024). *Världens mest protestantiska land.* Stolpe.

Brunkert, L. (2022). Overselling democracy–claiming legitimacy? The link between democratic pretention, notions of democracy and citizens' evaluations of regimes' democraticness. *Frontiers in Political Science, 4.*

Brunkert, L., Puranen B., Turska Kawa, A., & Welzel, C. (2024). Ideological polarization and crumbling institutional trust. Debunking the democracy falling stereotype (under publishing).

Carter, E. (2018). Right-wing extremism/radicalism: Reconstructing the concept. *Journal of Political Ideologies, 23*, 2.

Crum, B., & Oleart, A. (2023). *Populist parties and democratic resilience: A cross-national analysis of populist parties' impact on democratic pluralism in Europe.* Routledge Studies in Extremism and Democracy.

Czarnek, G., Szwed, P., & Kossowska, M. (2019). Right- and left-wing prejudice toward dissimilar groups in cultural and economic domains. *European Journal of Social Psychology, 49*(4), 807–823. https://doi.org/10.1002/ejsp.2548

Dalton, R. J., & Welzel, C. (Eds.) (2014). *The civic culture transformed.* Cambridge University Press.

Della Porta, D. (2008). Comparative analysis: Case-oriented versus variable-oriented research. In D. Della Porta, & M. Keating (Eds.), *Approaches and methodologies in the social sciences: A pluralist perspective* (pp. 198–222). Cambridge University Press; Cambridge Core. https://doi.org/10.1017/CBO9780511801938.012

Fassin, E. (2017). *Populisme. Le grand ressentiment.* Gallimard.

Foa, R. S., & Mounk, Y. (2017). The signs of deconsolidation. *Journal of Democracy, 28*(1), 5–15.

Gidron, N., & Hall, P. A. (2020). Populism as a problem of social integration. *Comparative Political Studies, 53*(7), 1027–1059. https://doi.org/10.1177/0010414019879947

Gessen, M. (2017). *The future is history: How totalitarianism reclaimed Russia.* Riverhead Books.

Goodhart, D. (2017). *The road to somewhere: The populist revolt and the future of politics.* Oxford University Press.

Government of Iceland, 2024–10–13.

Haerpfer, C., et al. (Eds.). (2018). *Democratization.* Oxford University Press.

Harrison, L. E., & Huntington, S. P. (2000). *Culture matters.* Basic Books.

Jungar, A. C., & Jupskås, A. R. (2014). Populist radical right parties in the Nordic region: A new and distinct party family? *Scandinavian Political Studies, 37*(3).

Kaltwasser, C. R., & Mudde, C. (2017). *Populism: A very short introduction.* Oxford University Press.
Kirsch, H., & Welzel, C. (2019). Democracy misunderstood: Authoritarian notions of democracy around the globe. *Social Forces, 98*(1).
Kuhnle, S., & Alestalo, M. (2017). *The modern Scandinavian welfare state.* Routledge.
Kuhnle, S., & Olsson, S. E. (2004). The developmental welfare state in Scandinavia, lessons for the developing world. UN Institute for Social Development. UNRISD, program papers.
Kvist, J., et al. (2012). *Changing social equality. The Nordic welfare model in the 21st century.* Bristol University Press.
Kvist, J., & Greve, B. (2011). Has the welfare model been transformed? *Social and Policy Administration,* 146–160.
Levitsky, S., & Ziblatt, D. (2018). *How democracies die.* Crown.
Lidegaard, B. (2011). *En fortaelling om danmark i det 20. Århundrede.*
Lindberg, S. I., Düpont, N., Higashijima, M., Kavasoglu, Y. B., Marquardt, K. L., Bernhard, M., Döring, H., Hicken, A., Laebens, M., Medzihorsky, J., Neundorf, A., Reuter, O. J., Ruth-Lovell, S., Weghorst, K. R., Wiesehomeier, N., Wright, J., Alizada, N., Bederke, P., Gastaldi, L., …, Seim, B. (2022). *V-Party Dataset v2.* V-Dem Institute. https://doi.org/10.23696/VPARTYDSV2
Lindroth, B. (2016). *Väljarnas hämnd, Populism och nationalism i Norden.* Carlssons förlag.
Lindroth, B. (2021). Vi som inte var med i kriget. Carlssons.
Lindroth, B. (2024). *Homo Affectus, Sverige, Norden och Folkhemstanken.* Atlas.
Mechkova, V., Lührmann, A., & Lindberg, S. I. (2017). How much democratic backsliding? *Journal of Democracy, 28*(4), 162–169.
Mounk, Y. (2018). *The people vs. democracy: Why our freedom is in danger and how to save it.* Harvard University Press.
Mudde, C. (2004). The populist Zeitgeist. *Government and Opposition, 39*(4), 541–563.
Mudde, C. (2017). *The populist radical right: A reader*. Routledge.
Mudde, C. (2019). *The far right today.* Medford Polity Press.
Norris, P. (2011). *Democratic deficits.* Cambridge University Press.
Norris, P. (2012). *Making democratic governance work. How regimes shape prosperity, welfare and peace.* Cambridge University Press.
Norris, P., & Inglehart, R., (2019), *Cultural backlash. Trump, Brexit, and Authoritarian populism.* Cambridge University Press.
Puranen, B. (2019). *Med migranternas röst. Den subjektiva integrationen,* Part 1, IFFS, 2019/2, part 1.
Puranen, B. (2020). What do migrants know about Sweden? *Kvartal.*
Puranen, B. (2021). *Med migranternas röst. Hur blir man värmlänning?* Part II, IFFS, 2021.
Puranen, B. (2022). När värdesystem kolliderar—utomeuropeiska migranters demokratisyn. In K. Abrahamsson et al. (Eds.), *Demokratin som bildningsväg.* Carlssons.
Puranen, B. (2023), *Med migranternas röst. Hur förändras värderingar?,* Part III, IFFS, 2023.
Putnam, R., (2007). E Pluribis Unum: Diversity and community in the 21st century. The 2006 Johan Skytte Price lecture. *Scandinavian Political Studies, 30*(2), 137–299.
Radkiewicz, P. (2017). Ideological inconsistencies on the left and right as a product of coherence of preferences for values. The case of Poland. *Polish Psychological Bulletin, 48*(1), 93–104. https://doi.org/10.1515/ppb-2017-0012
Rothstein, B. (2005). *Social traps and the problem of trust.* Cambridge University Press.
Rydgren, J. (2018). *The Oxford handbook of the radical right*. Oxford University Press.
Schäfer, A., Zürn, M., & Curtis, S. (2023). *The democratic regression: The political causes of authoritarian populism.* Polity Press.
SCB. (2022). Central statistics on voting results.
Snyder, T. (2018). *The road to unfreedom: Russia, Europe.* Tim Duggan Books.
SOU 2020:26. *En gemensam angelägenhet (A Common Concern)* part 1 and 2.
Statistics Iceland. *Innflytjendur 18.2% íbúa landsins,* 12 desember 2024. https://hagstofa.is/utgafur/frettasafn/mannfjoldi/mannfjoldi-eftir-bakgrunni-1-januar-2024/

Tolinson-Ting, K. (2020). *Social trust in Sweden: From contention to consensus.*
Welzel, C. (2013), *Freedom rising, human empowerment and the quest for emancipation.* Cambridge University Press.
Wikforss, Å. (2021). *Därför demokrati.* Fri tanke.

Bi Puranen is an associate professor in the History of Economics, Stockholm University and affiliated with the Institute for Future Studies, Stockholm Sweden. She is Secretary General of the World Values Survey and Research Associate in the EU Horizon-funded "TRUEDEM" project (Trust and Democracy). Earlier she was a professor of Future Studies and Communication Strategies at Sophia Antipolis, France. Her research focuses on democracy, migration, human security and changing values.

Lennart Joe Brunkert is a post-doctoral researcher at Leuphana University in Lüneberg, Germany and a Research Associate in the EU Horizon-funded "TRUEDEM" project. In his work, Brunkert analyzes questions of regime legitimation and public opinion from political science and social-psychology perspectives with a focus on quantitative measurement. Some of his recent publications have appeared in *Democratization, Frontiers of Political Science* and the *Journal of Democracy.*

Chapter 4
Welfare and Democracy in the Nordic Region—A Context for Policymaking

Haldor Byrkjeflot and Johan Strang

Abstract In this chapter, we discuss how arguments of democracy and democratization have been used to justify the welfare state in the Nordic region and how these arguments have changed through history. We start by discussing the social democratic account of the welfare state, according to which a foundational period in the 1930s lay the groundwork for the golden period of welfare state expansion in the region during the three decades after the Second World War. The welfare state was perceived as a crucial stage in a gradual democratization of Nordic society from the political realm, through the expansion of social welfare, to the ultimate establishment of economic democracy. As the mobilization for further expansion of democracy came to an end with the defeat for wage-earner funds in Sweden in the 1980s, the welfare state was transformed into an end in itself, and became a central focus for both the right and the left in Nordic politics. As the social democrats lost their hegemony, the welfare state came also to be seen as either the result of longer traditions of institutionalized associational life, or as the outcome of a specific Nordic individualist culture. Today, the connection between the welfare state and democracy is challenged by three forces: (1) the changing power balance between unions and employers, (2) the transformation of public participation in civil society, and (3) the cultural backlash towards post-material and post-family values in society. If the Nordic universal welfare state model continues to exist in the future, it will look very different from how it was originally defined during its foundational or golden periods.

Keywords Nordic model · Welfare states · Welfare democracy · Economic democracy · Social democracy

H. Byrkjeflot (✉)
University of Oslo, Oslo, Norway
e-mail: haldor.byrljeflot@sosgeo.uio.no

J. Strang
University of Helsinki, Helsinki, Finland

D. C. Nord (ed.), *Contemporary Nordic Public Policies*, Springer Polar Sciences,
https://doi.org/10.1007/978-3-032-11292-7_4

4.1 Introduction

The Scandinavian or Nordic Welfare State Model has been an entrenched term in international scholarly literature since Gøsta Esping-Andersen's path-breaking book *Three Worlds of Welfare Capitalism* was published in 1990. It has been propelled into political discourse and policymaking analysis both within and outside of the region. Indeed, today the welfare state has become one of the most predominant proofs of the "exceptionality" of the Nordic societies, and a central part of "the Nordic brand" that the Nordic countries use to promote themselves abroad. The Nordic Welfare State Model has also become a contested concept, which is claimed by various political actors from both the left and right. It is defined differently across various political and geographical settings—reflecting an "inner diversity" within the region (Esping-Andersen, 1990).

In this chapter, we will discuss how the arguments for democracy and democratization have been used to justify the further development of welfare arrangements in the Nordic region. We do not proceed from a particular definition of democracy—to the contrary, our argument is that our understanding of the central features of democracy itself has evolved along with changes in how the welfare state has been justified.

A defining feature of the Nordic Welfare State Model is the so-called principle of universalism. Emphasizing universalism was one of the ways in which Esping-Andersen, distinguished the Nordic model from the more mean-tested welfare state regime of the Anglo-Saxon countries and the more family and employer-based welfare state regime of Germany and France. Universalism referred to the idea of providing welfare programs to all members of society, regardless of their income, social status, family relations or other characteristics. To be sure, there are many exceptions to universalism in the Nordic countries today (see e.g. Kildal & Kuhnle, 2005), but in general the principle continues to have a stronger position in the Nordic countries than elsewhere. In addition to universality, Esping-Andersen argued that the Nordic Welfare State was comparatively generous in character, even to the extent that it has a redistributive ambition of increasing overall social and economic equality in society. The level of welfare services, such as pension, unemployment or sickness benefits, are generous enough to support a "de-commodified" or self-sustaining life when and if needed. The combination of universality and generosity is often said to be unique to the Nordic Welfare Model (Kangas & Kvist, 2018), as that the model is redistributive and based on the values of equality and solidarity (Graubard, 1986; Stråth, 2018).

Intra-Nordic comparisons reveal many differences between the institutional set-ups of the welfare state in the different Nordic countries, which are exemplified by policies, for example, related to labor relations, education and immigration. However, when compared more broadly to other countries from outside the region, these differences tend to become less significant making it possible to talk about a shared Nordic Model of Welfare. In looking for the reasons for the comparative similarities of welfare states in the region, it is crucial to note the role of Nordic cooperation, and the aspiration for convergence and harmonization that marked welfare policy discussions during substantial parts of the twentieth century (Kettunen et al., 2016).

Esping-Andersen put forward his definition of the Nordic Welfare State Model at the very end of what has later been characterized as "the golden period" of the welfare state. Globalization, Europeanization and neo-liberalization have transformed the Nordic welfare states and made them less unique from a European perspective. Indeed, by the 2020s, the Nordic welfare states have become less exceptional, less generous, and less equality-oriented in their traditional policy areas (pension, unemployment and sickness benefits) (e.g. Pedersen & Kuhnle, 2017). At the same time, the Nordic countries have arguably become *more* distinctive in terms of their actions related to gender equality (e.g. aims to enhance equal participation in work-life and politics) and family policies (family leave quotas and childcare). However, as there now seems to be less political interest in Nordic cooperation and harmonization in the welfare sector, there are also signs that the inner diversity among Nordic welfare states may be growing (Kettunen et al., 2016).

In this chapter, we focus on the relationship between the Nordic welfare state and democracy. Our argument is that while the welfare state has been seen historically as intimately connected to the concept of democracy, and largely perceived as a means for democratization of the Nordic societies, the criticism of the welfare state during the 1970s and 1990s—both from the left and the right—was also conducted in the name of democracy. During these periods, the welfare state was viewed as an obstacle to a "true" democracy in which individuals might have increased autonomy in making decisions over their own lives. Today, both Nordic democracy and the Nordic welfare state seem to be celebrated across political divides, yet the exact meaning of these concepts and the link between them seem less obvious (See Chap. 1 of this volume). The aim of our investigation of the historical relationship between welfare and democracy is to develop knowledge that may be relevant in the current effort to face the challenges related to the global erosion of democratic institutions and values. Knowledge of the past may also be relevant in understanding possible future combinations of welfare states and democracy.

The structure of the chapter is as follows: First, we present the social democratic account of the Nordic welfare state, from the foundational period in the 1930s, to its first period of expansion from 1945 to 1985. We then discuss examples of criticism of the social democratic welfare state from the 1970s onwards. We next examine the transformation of the Nordic welfare state during the 1980s and 1990s which not only redefined its character, but also gave rise to new scholarly perspectives on its historical origins and development. The chapter concludes with a discussion of the future direction of the Nordic welfare societies and their relation to the concept of democracy.

4.2 Social Democracy and the Nordic Welfare State

The traditional narrative of the origin and development of the Nordic welfare states is based on the dominant position of the Social Democratic parties in the region during much of the twentieth century. Sweden had a Social Democratic prime minister from 1932 to 1976 and the Norwegian Labour Party polled over 40 per cent of the votes in all elections during the same period. Social democracy was a bit weaker in Denmark and Finland, but, nonetheless, played a crucial role in national politics within both states. With their electoral success, the social democrats were positioned to take credit for the welfare model that emerged throughout the Nordic region.

The 1930s hold a special place in the traditional narratives of the Nordic welfare state. Whereas the political turmoil that followed the First World War and the Great Depression pushed many other European societies towards communism, fascism or Nazism, the Nordic countries opted for peaceful reform under Social Democratic leadership (Berman, 2006; Esping-Andersen, 1985). The idea was advanced that Social Democracy and welfare policy (*välfärdspolitik*) formed a bulwark against totalitarianism. This view was not merely constructed in hindsight. It was a key message of the Danish social democratic intellectual Hartvig Frisch as well as of the Swedish social scientist Gunnar Myrdal, who explained to an American audience that progressive social reforms were necessary in order to safeguard democracy "with dictators as neighbors" (Frisch, 1933; Kurunmäki & Strang, 2010; Myrdal, 1939).

A crucial element of the Nordic labor movement in the 1930s was its abandonment of class struggle in favor of a more pragmatic approach to public policy that embraced elements of both capitalism and nationalism (Berman, 2006; Esping-Andersen, 1985; Stråth, 2018). The 1930s saw the Social Democrats facilitating class compromises between the employers and unions within the labor market, and between the workers and the farmers in parliament. This created a virtuous circle between labor, capital and producers and between socialism, capitalism and democracy (Kettunen, 2019). These policy actions toward the labor market earned the attention of politicians and observers from abroad, including President Franklin Roosevelt and the American journalist Marquis Childs who first launched the idea of a Nordic "middle way" (Childs, 1936; Rom-Jensen, 2017).

Another tension that characterized Nordic social democracy in the 1930s was one between progressivism and nationalism. On the one hand, the 1930s was a decade when leading Social Democrats embraced progressive, rational and even technocratic ideas, epitomized by the Swedish intellectuals Alva and Gunnar Myrdal who argued for welfare reforms by means of utopian, yet modernistic and highly rational argumentation. Housing, education, health, and reproduction were, they suggested, issues that had to be approached scientifically, "with the romanticism of an engineer" (Myrdal, 1932, p. 25).

On the other hand, the 1930s was also the decade when Social Democracy embraced themes of nationalism and a rhetoric focused on needs of "the people." In Sweden, Prime Minister Per-Albin Hansson adopted the concept of a "people's home"—*folkhemmet*—from the national conservatives and combined it with his push for comprehensive welfare reforms. Likewise, in Denmark, the Social Democratic party program of 1934 proclaimed "*Danmark for folket*" (Denmark for the people). In Norway, the Labor Party slogan for the 1933 elections was "*Hele folket i arbeid*" (The whole people at work).

At the same time, myths about the free Nordic peasant (Koht, 1926) and age-old traditions of democracy (Lindahl, 1981) were also promoted and made part of the social democratic narrative of the Nordic nations. These narratives gave workers and farmers the historical role of fostering and expanding Nordic democracy (Kurunmäki & Strang, 2010). In a way, both democracy and progressivism were turned into essential parts of the national legacies, which arguably made the space for totalitarian rhetoric smaller in Scandinavia than in many other parts of Europe (Nielsen, 2009).

The idea that progressive social democratic public policies had saved the Nordic countries from totalitarianism was also crucial in establishing the path that the Nordic countries would pursue after the Second World War. The notion of a Nordic "consensual model" of democracy, based on large public policy compromises gained wide currency and became a central part of the ideological positioning of the Nordic countries during the Cold War. While these nations clearly belonged to the realm of western capitalist democracies, the rhetoric of a "middle way," combining the best parts of capitalism and socialism, "freedom" with "welfare," were crucial parts in the legitimation of a distinctive Nordic social model during this period (Kettunen & Petersen, 2022; Kurunmäki & Strang, 2010; Nelson, 1953).

During the post World War II period, democracy was a key concept that both the West and the East tried to monopolize for their own purposes. However, the idea that the Nordic countries were truly exceptional in this regard was particularly anchored in the existence of the social democratic welfare state. The association of welfare with democracy became a leading theme of several postwar books on the Nordic countries that received wide attention. Whereas some of these books tended to frame the Nordic social model as a "middle way" in the ideological Cold War, others presented an image of the Nordic countries as a *maximal* democracy, combining advanced representative democracy with the democratization of a range of institutional spheres such as industry, schools, culture, political parties, and cooperatives (Byrkjeflot, 2001). The idea was that societies of the region were becoming increasingly democratized as the welfare state expanded.

The social democratic vision was that democracy and welfare reforms were mutually reinforcing. However, It was not always clear whether the welfare state was the end-goal in itself. Instead, the welfare state was often projected as only a step towards a fuller democratization of society (Esping-Andersen, 1985; Friberg, 2013). According to this social democratic narrative, the struggle among progressive forces to cultivate markets and to develop advanced democratic societies of the future would result in three transformations. The first transformation had already occurred. This was when the working class and farmer alliances brought social democrats into government. The second transformation was to be the ongoing development of the social welfare state. The third transformation would witness industrial and economic democracy. Exactly what was meant by industrial and economic democracy varied somewhat among the Nordics and also over time. However, the central idea was to give workers more power over their working conditions and eventually arrive at worker control over some of the profits associated with private business.

Yet as the welfare state expanded over the subsequent decades and employers resisted the idea of an economic democracy, the welfare state, itself, became *the* central focus for much of the political discourse related to democratic reform in the Nordic societies. Before examining this phenomenon, however, we will first investigate how the welfare state became an object for criticism both from the political right and left and then discuss the democratic implications of these critiques.

4.3 1970–1990: Criticism of the Social Welfare State from the Left and the Right

There had, of course, been longstanding criticism from the political right regarding the social democratic ambitions for expanding the welfare state. Particularly during the so-called "planned economy debates" of the post-war World War II period, many Nordic liberals and conservatives argued that the social democratic welfare state had gone too far in restricting the lives and freedom of its citizens (Sejersted, 2005: 326–329). However, this original criticism was further propelled by neoliberal activism from the late 1970s onwards. The character of the social democratic welfare state had already been subject to equally strong, and partly similar, criticism from the new left.

From the late 1960s, a radicalized left wing in Scandinavia had argued that the Nordic welfare state had become an apparatus that merely masked the misfortunes of capitalism. As such, it had failed to live up to its promises of providing a truly democratic and socialist society. The time was ripe, these critics argued, to take the third step towards industrial and economic democracy. This leftwing critique encouraged the Nordic social democrats to develop programs that became more radical in their ambition for justice and equality, most notably the wage-earner funds developed by Swedish Social Democrats and the unions in the 1970s (Arvidsson & Bentson, 1980; Lingås, 1970). Simultaneous with these efforts towards economic democracy, were other debates on the left about bringing democracy closer to the ordinary lives of people (Køber, 2017). There were calls for a reform movement related to codetermination in working-life (Gustavsen, 1981). Crucially, this leftwing criticism was directed at a welfare state, which they viewed as being too bureaucratic, authoritarian and not delivering on promises of democratic participation.

Arguably, such criticism contributed to the subsequent rise of the new neoliberal right, which recirculated many of the same arguments pertaining to the broken relationship between democracy and welfare. From the perspective of the right, the social democratic welfare state was viewed as being excessively paternalistic. A common argument was made that the welfare state had given rise to a new governing class of social experts who interfered too much in the lives of the individual citizen. The critique against what they perceived to be a paternalist, even interventionist colonization of the private sphere, became the favorite theme of many right-leaning politicians and public intellectuals from the 1970s to 1990s in the Nordic region (Dich, 1973; Fonsmark, 1990; Rasmussen, 1991, 2017).

The Nordic political right also criticized the existing welfare state by calling for freedom of individual choice and a more consumerist notion of democracy (Olsen, 2018). These critics also promoted a more restricted legalistic view of democracy based around the idea of property and contract rights (Fonsmark, 1990; Myhrman et al., 1987; see Strang, 2025). According to their view, in order to move forward toward a more democratic way of organizing the community, it would be necessary to rely more on civil society and private initiatives. They also called for the scaling down of the welfare apparatus in order to avoid overburdening the population with further tax increases.

Arguably, it was this criticism from the right that proved more decisive in changing the direction of the Nordic Welfare State Model. The leftwing movement for a further democratization of society stalled in the late 1970s and in the early 1980s. A counter-mobilization from employers' organizations and political parties on the right then became more prominent and decisive in public policy formulation.

This ideational shift also affected the Social Democrats themselves. The Norwegian Labor Party's Congress in 1981 altered its ambitions for a transition into socialism (Johansen, 1995) and abandoned its critical attitude towards private property and managerial prerogatives. The same trend was largely seen in Sweden when the wage earners funds were ultimately abolished in 1991. Similar developments took place in Finland (Kärrylä, 2021) and Denmark (Køber, 2024; Toubøl & Gielfeldt, 2013). In the 1990s, Social Democrats across the Nordic region adopted new policies aimed at reinventing themselves in an age of globalization and Europeanization.

Interestingly, it was also in the late 1980s and early 1990s, when the Nordic left abandoned its ambitions for moving to a more genuine and thorough democratization of society. As a consequence, the welfare state became canonized as "the jewel in the crown" of the Nordic community and became the end-point for the social democratic "project" within the region. As Finland and Sweden were hit by a hard recession in the early 1990s, drastic cuts were made to their welfare provisions. From then on, it was no longer the *expansion*, but rather the *defense* of the welfare state that became the key aim of social democracy.

Soon the debate became no longer one of being for or against the welfare state, but rather each contesting group tried to claim and define the Nordic welfare state for its own political purposes. Conservatives continued to claim that it was necessary to reduce state influence in welfare policy, but they now distinguished between the old, cold-hearted bureaucratic social democratic welfare *state*, on the one hand, and their own vision of a warmer and more welcoming welfare *society*, on the other. From their point of view, Nordic citizens should be able to distinguish more clearly between state responsibilities and voluntary efforts to encourage private initiatives and thus avoid overloading state budgets. The political left, as will be discussed later, came up with its own interpretations.

As the welfare state emerged as "the jewel in the crown" of Nordic societies, almost all political parties became its supporter, and new interpretations were introduced that could provide alternatives to the social democratic hegemony in the legitimation of the welfare state.

4.4 New Interpretations: The Nordic Model as Associative Governance or as an Expression of Nordic Culture

Interestingly, as the golden period of the welfare state was reaching its endpoint, there emerged new interpretations that raised questions regarding the centrality of Social Democracy in the history and development of the Nordic welfare states. The American historian Peter Baldwin (1989), for example, has pointed to the fact that many welfare reforms were originally introduced with substantial agreement across political divides, and that many features associated with the Nordic welfare model actually predate the social democratic era. Indeed, it was undoubtedly correct to note that universal welfare schemes had been supported by, for example, Danish farmers who wanted the state to share the social costs associated with keeping labor on their farms. Similarly, health, accident and disability insurances had been introduced at the turn of the century with inspiration from Bismarck's Germany, although they were translated into more universal schemes when imported to the Nordics (see e.g. Kildal & Kuhnle, 2005).

In this new interpretation, the defining feature of the Nordic model has become no longer only the balance of power between employers and employees, or the strong position of social democracy, but rather the many negotiated compromises between a wide diversity of interests in the Nordic societies. Some historians argued that the roots of the Nordic Model should be actually traced back to the cooperative movements and associations that were key during the nation-building period within the region during the nineteenth century (Alapuro & Stenius, 2010; Hilson, 2008). Other social scientists, in turn, have alternatively identified the Nordic Model with overlapping conflict cleavages in society (Rokkan et al., 1981; Sass & Kuhnle, 2022) an institutionalized consensual style of politics (Kuhnle, 2009; Rothstein, 2011); or a specifically Nordic kind of state-friendly civil sphere that created a high degree of societal trust (Engelstad & Larsen, 2020; Kuhnle & Selle, 1992).

In parallel to these types of interpretations, there emerged in the 1990s a cultural criticism of the social democratic monopoly of the Nordic welfare state. It was now argued that the roots of the Nordic Model were found, neither in social democracy nor in nineteenth century popular movements, but in the traditions of the free Nordic peasant or in the Reformation and the influence of the state-church (Sørensen & Stråth, 1997). With regard to the latter, it has been argued that the marriage of state and church facilitated various forms of local administration and democracy and accompanying public social responsibility. The merged secular and religious norms gave impetus to an idea of the good and strong state. Consequently, the Nordic welfare state was seen not as much the result of democratic socialism as it was of secularized Lutheranism. According to this interpretation, it was Luther rather than Marx that was the real inspiration for the Nordic welfare states (Nelson, 2017; Stenius, 2010).

Even more radically, it has been suggested by some scholars that the Nordic Model was not so much a political creation but an outcome of specific Swedish or Nordic cultures (Berggren and Trägårdh 2011). According to Berggren and Trägårdh (2006, 2022), the Nordic welfare state was built on a contract between the state and the individual, whereas other relationships, such as family ties or intermediate organizations, played a much less significant role. They refer to this arrangement as "statist individualism." Notwithstanding the criticism that this notion of statist individualism overlooks the importance of the people's movements and associational life in the Nordic region, as well as the central role of family policies in the Nordic welfare state, Berggren and Trägårdh has provided a fresh perspective that has provoked interesting political debates.

At the same time, such cultural criticisms of the social democratic narrative has also contributed to a gradual de-politization of the Nordic welfare model and to the disassociation of the welfare state from not merely social democracy, but also democracy itself. If the Nordic welfare state could be explained by referring to cultural heritage, and if it could be maintained by further development of an individual contract between citizens and the state, then there was conceivably less need for democratic politics.

The de-politization of the Nordic Model has been not without political consequences. In 2008, Mary Hilson noticed the uncomfortable emergence of cultural explanations for the welfare state with the rise of xenophobic welfare populism in the Nordic region (Hilson, 2008: 175–176). If the Nordic welfare state was truly an expression of a specific Nordic culture, it could become quite easy also to argue that "foreigners" are inherently incapable of understanding this exceptional Nordic welfare culture (Bell et al., 2023). Such an interpretation might encourage a major rethinking of migration and multiculturalism policies within in each of the Nordic countries.

Moreover, it is also worth pondering whether the cultural and historical interpretations of the welfare state contributed to the belief that the Nordic Model was a fixed form that could withstand any challenges from marketization and neoliberalism. If the Nordic Model is cemented in cultural and historical legacies beyond social democracy, it might withstand economic changes, cuts and privatizations. Indeed, at the 2011 World Economic Forum in Davos the Nordic Model was presented as both culturally anchored in specific Nordic values and packaged as a brand that could be used for marketing purposes (Berggren & Trägårdh, 2006, 2011). Similarly, the *Economist* presented the Nordic Model in 2013 as a blueprint for how to reform the public sector, at the same time making the state more efficient and responsive (The Economist, 2013; Byrkjeflot et al., 2022). It was at this point that it became very clear that "Nordic" had replaced "social democracy" as the main antecedent of the welfare model and that a more individualistic and "free choice" idea of welfare democracy had become increasingly evident in its presentation (Byrkjeflot, 2024; Kuhnle, 2009).

4.5 Current Challenges to Nordic Welfare Democracy

In recent decades, the Nordic Welfare State Model has been subject to both political and scholarly contestation. The traditional interpretation, according to which the welfare state was a stage in the democratization of the Nordic societies, has been replaced by a struggle between those who see it as the endpoint of a social democratic project, and those who view it as the result of older traditions of associational life, or even as the outcome of a specific Nordic individualist culture. These different interpretations emphasize either sociopolitical aspects, the role of civil society and people's movements, or the influence of religion, state building and long-standing peasant traditions. To conclude the chapter, we want to look forward and consider the current and future challenges faced by the Nordic Welfare State Model and, in particular, its relation to democracy in the region. We will discuss these challenges in relation to (1) the decline of social democracy, (2) the transformation of associational life, and (3) a possible cultural shift towards liberalism and conservatism in Nordic societies and the emergence of the "the silent revolution".

4.5.1 A Changing Power Balance Between Employers' Associations and Unions, and the Decline of Social Democracy

Today, social democracy is in crisis throughout Europe. In the Nordic region, the Social Democratic parties remain comparatively strong, even if they have declined considerably in electoral support. While the Nordic welfare state has become a source of regional pride, the Social Democrats have lost their hegemony over its history and meaning. Leftwing parties have claimed its redistributive agenda, while the neoliberal right has tried to monopolize its economic competitiveness dimension. Where the Liberals and Greens pose as true Nordic progressives, the populist national conservatives of the region capitalize on the nationalist, even nostalgic, "people's home" rhetoric. To whatever extent there will be a Nordic Welfare Model in the future, it is likely to be less closely associated with social democracy, both as a political party and as a political ideology.

The relative decline in union power in relation to employers in the Nordic countries is also a major challenge to the established welfare regime. Since 1985, there has been a fall in trade union membership in Finland, Sweden and Denmark (but less dramatically in Norway). State intervention in wage negotiations and a move towards decentralized and more individualized bargaining are other developments that have challenged the established tripartite model. For example, during the spring of 2024, Finland was struck by a wave of political strikes where the unions were protesting against the conservative government. With Finnish employers withdrawn from the tripartite negotiations, the government referred to their "democratic" mandate to pursue reductions in sickness, unemployment and housing benefits through legislation (Bender & Kjellberg, 2024). Democracy, it seems, is no longer conceived of in terms of labor market negotiations, but rather something that was manifested in, and monopolized by, a parliamentary majority.

The broader Nordic institutional system of a tripartite bargaining structure, such as the national mediator's office, arbitration courts and technical calculation committees, are all put under stress in such an increasingly polarized political environment. While the Nordic welfare state system still relies on a tripartite governance model, this arrangement is also potentially challenged by new EU legislation which allows for the state to regulate a minimum wage (Dølvik, 2022; Kjellberg, 2023).

4.5.2 Decline of Public Participation in Spheres of Welfare and Civil Society

Another key challenge to the Nordic Welfare State Model is the withering away of traditional "people's movements" and voluntary organizations. This leads to a decline in the interaction between grassroots in civil society, political parties and the state at both local and national levels. Instead of acting in a traditional corporatist manner to shape public policy, the latest generation of associations often explicitly challenges the work of national policymakers through political protest, direct lobbying, or social media action. The decline of traditional Nordic democratic neo-corporatism, with a whole range of organizations bringing their influence to bear in the political system, is also reflected in how these organizations now use domestic and international courts to further their interests. They often refer to international norms and to security and environmental threats that have to be dealt with at global level in order to justify their positions (Schaffer, 2019; Schaffer & Schoultz, 2024).

To be sure, recent surveys indicate that the Nordic states continue to have a vibrant civil society and that many of the institutions related to neo-corporatism still exist (Christiansen, 2018; Henriksen et al., 2019). The change in institutionalized practices and associative relations seem to reflect a gradual rather than definitive trend. Indeed, there are still a variety of modes of governance coexisting in Nordic societies today, such as cooperatives, housing associations, collegial organizational forms and foundations (Dragsted, 2021; Køber, 2024; Sahlin & Eriksson-Zetterquist, 2024). These forms may be revitalized and democratized under given circumstances and thereby contribute to upholding the participatory traditions of Nordic societies. It may also be argued that these governance forms have been embedded in a particular national culture. We now turn to explore this kind of argument.

4.5.3 *Long Term Development in Culture: Is There a Backlash to the "Silent Revolution"?*

A central feature of the Nordic welfare model has also been its progressive and liberal ethos of emancipating individuals from their dependence not merely from the dominance of the marketplace or big capital, but also from patriarchal structures associated with the family, religion or other representatives of conservative morality. This was the focus of Trägårdh's concept of statist individualism. However, as observed by Inglehart and Norris (2017) there may be signs of a reversal of these trends and a "backlash" away from progressive values towards more authoritarian and anti-pluralist thinking, which may partly explain the rise and continuous strength of the national conservative populist parties across the West, and in the Nordic countries (Bell et al., 2023). There is now significant support in the Nordic electorate for the idea that non-citizens should not receive the same level of welfare benefits as citizens and an even stronger skepticism about providing dissimilar cultural groups with full welfare rights. As mainstream parties adapt to the right-wing populist parties' positions, this might be regarded as another challenge to the universal welfare state tradition of the Nordic countries.

The extent to which these developments signal a true "backlash" to the Nordic welfare democratic model, however, is debatable. To be sure, the various national conservative groups often seem to express a nostalgia for the past and "the traditional family values" that the welfare state supposedly has eradicated. At times, they also refer to feminism and LGBT-issues as part of their anti-immigration agendas. The center-right parties, in general, push for free choice of welfare providers and in outsourcing of welfare services to private providers. The social democratic parties argue for a social investment strategy, which aims at long-term social returns, such as developing a more qualified workforce or improving health outcomes It remains to be seen whether the Nordic countries can find a way of balancing individual choice and social investment in a way that does not undermine the democratic ambitions associated with the welfare state.

4.6 Prospects for Future Development of Welfare Democracy: A Few Concluding Words

This chapter has focused on how ideas related to the Nordics as welfare states or "welfare democracies" have emerged and developed, as well as on how they continue to influence the political debates in the region. The aim has been to explore and discuss the implicit and explicit relationship between the Nordic welfare state and democracy. In this manner a conceptual context is provided for the discussion of specific public policy efforts later in this volume.

We have outlined the social democratic narrative of the Nordic models of welfare and democracy and have discussed how this has been challenged by alternative associative governance and culturalist narratives. According to the social democratic viewpoint, it is the balance between labor and employers and the democratic ambition behind and beyond the welfare state that has been determinative in accounting for the rise of the Nordic welfare state. According to this perspective, if the state and employers get too powerful in comparison with labor and unions then the universal welfare state and, in particular, its democratic aspects are likely to be undermined.

The action emphasized in the second narrative—associative governance—is situated more in the broader sphere of civil society and the relationship between various associations, and the state. This perspective poses less of a direct challenge to the social democratic narrative, but places more emphasis on the broader coalition of people's movements and political parties in ensuring democracy. The social democratic narrative is more explicitly focused on the critical importance of labor-employer relations.

The third narrative, the cultural construction approach, alternatively, is more attuned to the rise of welfare rights and the more recent contract between the individual and the state. From such a point of view what matters most for democratic development is the menu of welfare rights offered and the possibility to exercise freedom of choice in welfare provision. It is therefore a more individualistic perception of democracy. Whereas the social democratic and cultural construction perspectives focus on the economy and culture, respectively, the associative governance narrative seeks to combine the economic and cultural aspects of development (Byrkjeflot, 2024).

Recent trends associated with each of these three narratives have been examined, such as the changing power relations between employers and unions, the transformation of civil society, as well as the neo-populist "backlash" These trends provide a challenge to both the traditional democratic architecture of Nordic welfare states as well as the more recent combination of individualistic and traditionalist welfare measures. In order to avoid a further move towards a welfare technocracy within the Nordic states, it will be useful to learn from the experiences of past mobilizations around welfare where voluntary measures helped build the welfare system from below. There is a need to investigate further how the shifting architecture of welfare states relates to ideas of democracy, in general, as well as the specific democratic aspects of the Nordic welfare state.

As we are moving towards a possible post-global era (see Chap. 2 in this book), it may become more difficult to negotiate common solutions for the long-term development of welfare states in Europe. Furthermore, as challenges like war and climate change are dominating the discourse, surveys show that there has been a shift in public opinion towards pessimism related to whether basic welfare needs will be taken care of in the future.[1] Likewise, both the internal and external perceptions of the Nordics as progressive welfare states may be challenged, and as a consequence, it may become increasingly difficult to argue that the current societal architecture is particularly democratic. However, the combination of de-globalization efforts

[1] A survey from Kantar in April 2024 show that 7 out of 10 fear that the quality of health care services in Norway will be in decline in the future.

and increased politization may also provide an opportunity for revitalizing some of the institutions that have been associated with Nordic democracy. In particular, the voluntary, membership-based organizations in the welfare field may be revived somewhat.

Although the Nordic societies are sufficiently similar that it is possible to speak of a Nordic Model, there are still specific aspects of it that are not shared by all the Nordic countries. Some degree of inner diversity continues to exist. Yet, there is an increasing tendency throughout the region to associate the welfare state with education, health and family policies rather than with labor market relations. Also, in common, is the increased presence of private actors in various fields of the welfare state (see Chap. 5 in this volume). However, it seems as if Sweden has progressed furthest down the path from an ideologically motivated social democracy in the 1970s to an ideologically motivated neoliberal competitive state today (Andersson, 2025). The Danish, Norwegian and Finnish welfare states were always more divided between public, private and third sector than Sweden, and continue to be so today. Norway in turn, which has been somewhat protected from such policy reforms due to its oil wealth and a relatively active civil society, has nonetheless seen a trend toward for-profit privatization of childcare (Byrkjeflot forthcoming).

In conclusion, the Nordic welfare state model still exists but it has become more oriented toward family policy and gender equality concerns. As it has expanded into new areas and arenas, it has become more diversified, complex and, perhaps, even more technocratic. As a consequence, the Nordic welfare state, within its existing architectural structure, has had a hard time fulfilling the early social-democratic ambition of the welfare state as a means to achieve the democratization of society.

References

Alapuro, R., & Stenius, H. (2010). *Nordic associations in a European perspective.* Nomos.

Andersson, J. (2025). Capital interest: Privatisation in Sweden. In Andersson, & Howell (Eds.), *Neoliberalism in the Nordics* (pp. 96–113). Routledge.

Arvidsson, H., & Bentson, L. (1980). *Makten, socialismen och Demokratin: Om det förstatligade samhället.* Zenit.

Bell, D. A., Valenta, M., & Strabac, Z. (2023). Nordic welfare chauvinism: A comparative study of welfare chauvinism in Sweden, Norway and Finland. *International Social Work, 66*(6), 1786–1802.

Bender, G., & Kjellberg, A. (2024). Finland rattled by political strikes against the conservative governments' anti-union agenda. Onlabor.org. Downloaded 25.6.2024. https://onlabor.org/finland-rattled-by-political-strikes-against-the-conservative-governments-anti-union-agenda/

Berggren, H., & Trägårdh (2006). *Är Svensken Människa?* (Is the Swede a human being?) Norstedts.

Berggren, H., & Trägårdh, L. (2011) Social trust and radical individualism: The paradox at the heart of Nordic capitalism. In *The Nordic way*, Davos. Global Utmaning.

Berggren, H., & Trägårdh, L. (2022). *The Swedish theory of love: Individualism and social trust in modern Sweden.* University of Washington Press.

Berman, S. (2006). *The primacy of politics: Social democracy and the making of Europe's twentieth century.* Cambridge University Press.

Byrkjeflot, H. (2001). The Nordic model of democracy and management. In Byrkjeflot et al. (Eds.), *The democratic challenge to capitalism: Management and democracy in the Nordic countries* (pp. 19–45). Fagbokforlaget.
Byrkjeflot, H. (2024) A Nordic model of associative governance. In A. B. Lund, H. Byrkjeflot, & S. Christensen (Eds.), *Associative governance in Scandinavia: Organizing societies by combining together*. Routledge.
Byrkjeflot, H. (forthcoming). Welfare states and civil society in the Nordic region. In B. Greve (Ed.), *Handbook on social policy and civil society.*
Byrkjeflot, H., Mjøset, L., Mordhorst, M., & Petersen, K. (Eds.) (2022). *The making and circulation of Nordic models, ideas and images. Nordic studies in a global context*. Routledge.
Childs, M. W. (1936). *Sweden: The middle way.* Yale University Press.
Christiansen, P. M. (2018). Still the corporatist darlings? In P. Nedergaard, & A. Wivel (Eds.), *The Routledge handbook of Scandinavian politics* (pp. 36–48). Routledge
Dich, J. (1973). *Den herskende klasse. En kritisk analyse af social udbytning og midlerne imod den.* Borgen.
Dragsted, P. (2021). *Nordisk socialisme.* Gyldendal
Dølvik, J. E. (2022). Strengthening the Nordic working life model–A precondition for successful transition to the future of work. Background paper for labor minister meeting. FAFO paper 2022:13.
Engelstad, F., & Larsen, H. (2020). Nordic civil spheres and pro-civil states. In J. C. Alexander, L. Anna, & A. Voyer (Eds.), *The Nordic civil sphere* (pp. 39–63). John Wiley & Sons.
Esping-Andersen, G. (1985). *Politics against markets: The social democratic road to power.* Princeton University Press.
Esping-Andersen, G. (1990). *Three worlds of welfare capitalism.* Princeton University Press.
Fonsmark, H. (1990). Historien om den danske utopi. Gyldendal.
Friberg, A. (2013). *Demokrati bortom Politiken* Atlas.
Frisch, H. (1933). Pest over Europa. Bolschevisme—Fascisme—Nazisme. Henrik Koppels Forlag, København.
Graubard, S. R. (Ed.). (1986). *Norden: The passion for equality.* Oxford University Press.
Gustavsen, B. (1981). Nordic democracy. In E. Allardt et al. (Eds.), *Industrial democracy.* Det Danske Selskab.
Henriksen, L. S., Strømsnes, K., & Svedberg, L. (2019). Comparative and theoretical lessons from the Scandinavian case. In L. Henriksen, K. Strømsnes, & L. Svedberg (Eds.), *Civic engagement in Scandinavia: Volunteering, informal help and giving in Denmark, Norway and Sweden* (pp. 199–207).
Hilson, M. (2008). *The Nordic model: Scandinavia since 1945.* Reaktion Books.
Inglehart, R., & Norris, P. (2017). Trump and the populist authoritarian parties: The silent revolution in reverse. *Perspectives on Politics, 15*(2), 443–454.
Johansen, T. A. (1995) Bedriftsøkonomisk utvikling i en økonomisk krisetid *AHS-serie* A 1995–2.
Kangas, O., & Kvist, J. (2018). Nordic welfare states. In *Routledge handbook of the welfare state* (pp. 124–136). Routledge.
Kärrylä, I. (2021). *Democracy and the economy in Finland and Sweden since 1960: A Nordic perspective on neoliberalism.* Springer Nature.
Kettunen, P. (2019). The concept of society in the making of the Nordic welfare state. In S. Kuhnle, P. Selle, & S. E. Hort, (Eds.), *Globalizing welfare: An evolving Asian-European dialogue* (pp. 143–161). Edward Elgar Publishing.
Kettunen, P., & Petersen, K. (2022). Images of the Nordic welfare model: Historical layers and ambiguities. In *The making and circulation of Nordic models, ideas and images* (pp. 13–33). Routledge.
Kettunen, P., Urban Lundberg, U., Österberg, M., & Petersen, K. (2016). The Nordic model and the rise and fall of Nordic cooperation. In J. Strang (Ed.), *Nordic cooperation: A European region in transition* (pp. 69–91). Routledge.

Kildal, N., & Kuhnle, S. (2005). The Nordic welfare model and the idea of universalism. In Kildal & Kuhnle (Eds.), *Normative foundations of the welfare state: The Nordic experience* (pp. 13–33). Routledge.

Kjellberg, A. (2023). The Nordic model of industrial relations: Comparing Denmark, Finland, Norway and Sweden (Cologne: the New Trends and Challenges in Nordic Industrial Relations conference).

Koht, H. (1926). *Norsk Bondereising. Fyrebuing til bondepolikken.* Aschehoug.

Kuhnle, S. (2009). The Nordic model: Ambiguous, but useful concept. In H. Obinger, & E. Rieger (Hg.) *Wohlfahrtsstaatlichkeit in entwickelten Demokratien: Herausforderungen, reformen und perspektiven* (pp. 275–294). Campus Verlag.

Kuhnle, S., & Selle, P. (Eds.). (1992). *Government and voluntary organizations.* Avebury, Ashgate Publishing Ltd.

Kurunmäki, J., & Strang, J. (2010). Nordic democracy in a world of tensions. In J. Strang, & J. Kurunmäki (Eds.), *Rhetorics of Nordic democracy. Studia Fennica Historica* 17, Finnish Literature Society.

Køber, J. V. (2017) Et spørgsmål om nærhed—Nærdemokratibegrebets historie i 1970'ernes Danmark, Ph.D. afhandling, Saxo-Instituttet, Københavns Universitet.

Køber, J. V. (2024). Demokratisering af økonomien i det 21. århundrede. *Agora, 41*(2–3), 179–200.

Lindahl, S. (1981). Early democratic traditions in the Nordic countries. In E. Allardt (Ed.), *Nordic democracy: Ideas, issues and institutions* (pp. 15–43). Danske.

Lingås, L (ed. 1970) *Myten om velferdsstaten* Oslo, Pax

Myhrman, J., Strömholm, S., & Petrén, G. (1987). *Marknadsekonomins rättsliga grundvalar.* Timbro.

Myrdal, G. (1932). Socialpolitiskt dilemma II. *Spektrum, 2*(4), 13–31.

Myrdal, G. (1939). Maintaining democracy in Sweden, I. With dictators as neighbors, II. The Defence of Democracy. A Bonnier reprint from Survey Graphic—Magazine of Social Interpretation, May–June, 4.

Nelson, G. R. (Ed.) (1953). *Freedom and welfare. Social patterns in the Northern countries of Europe.* The Ministry of Social Affairs of Denmark, Finland, Iceland, Norway and Sweden.

Nelson, R. H. (2017). *Lutheranism and the Nordic spirit of social democracy: A different protestant ethic.* Aarhus Universitetsforlag.

Nielsen, N. K. (2009). *Bonde, stat og hjem. Nordisk demokrati og nationalisme-fra pietismen til 2. Verdenskrig.* Aarhus universitetsforlag.

Olsen, N. (2018), Palgrave Macmillan, Cham

Pedersen, A. W., & Kuhnle, S. (2017). The concept of a 'Nordic model'. *The Nordic Models in Political Science: Challenged, but Still Viable*, 219–238. Fagbokforlaget

Rasmussen, A. F. (1991/2017). *Fra socialstat til minimalstat.* Lindhardt og Ringhof.

Rom-Jensen, B. (2017). A model of social security? The political usage of Scandinavia in Roosevelt's new deal. *Scandinavian Journal of History, 42*(4), 363–388.

Rokkan, S., et al. (1981). The growth and structuring of mass politics. In E. Allardt (Ed.), *Nordic democracy* (pp. 53–79). Det Danske Selskab.

Rothstein, B. (2011). *The quality of government: Corruption, social trust, and inequality in international perspective.* University of Chicago Press.

Sahlin, K., & Eriksson-Zetterquist, U. (2024). *Revitalizing collegiality: Restoring faculty authority in universities* (p. 276). Emerald Publishing.

Sass, K., & Kuhnle, S. (2022). *The gender cleavage: Updating Rokkanian theory for the twenty-first century.* International Studies in Gender, State & Society.

Schaffer, J. K. (2019). How democracy promotion became a key aim of Sweden's development aid policy. In K. Bjørkdahl, & A. B. Puyvallé (Eds.), *Do-gooders at the end of aid* (pp. 143–170). Cambridge University Press.

Schaffer, J. K., & Schoultz, I. (2024). Introduction: Legal mobilization in Nordic civil society. *Nordic Journal of Human Rights, 42*(1), 1–10.

Sejersted, F. (2005) Sosialdemokratiets tidsalder. Norge og Sverige i det 20. Århundre Oslo: Pax

Stenius, H. (2010). Nordic associational life in a European and an inter-Nordic perspective. In R. Alapuro, & H. Stenius (Eds.), *Nordic associations in a European perspective: European civil society (8). Nomos* (pp. 29–86).

Strang, J. (2025). The neoliberal mobilization of law, constitutionalism and individual rights in Sweden 1980–2000. In J. Andersson, & C. Howell (Eds.), *Neoliberalism in the Nordics.* Routledge.

Stråth, B. (2018). The cultural construction of equality in Norden. In S. Bendixsen, M. B. Bringslid, & H. Vike (Eds.), *Egalitarianism in Scandinavia: Historical and contemporary perspectives* (pp. 47–64).

Sørensen, Ø., & Stråth, B. (Eds.). (1997). *The cultural construction of Norden.* Scandinavian University Press.

The Economist 2.2. (2013). The Nordic model, the Next Supermodel.

Toubøl, J., & Gielfeldt, J. K. (2013). Den fejlslagne kampagne for økonomisk demokrati som faktor i Arbejderbevægelsens politiske udvikling *Arbejderhistorie. Tidsskrift for historie, kultur og politik*, bind 2013, nr. 3, 50–71.

Haldor Byrkjeflot is professor of sociology at the University of Oslo. He was the academic director of UiO: Nordic (2015–2019) as well as chair of the board (2018–2019) of ReNEW a university hub studying the Nordic region. Byrkjeflot's publications are in the fields of historical comparative studies, public sector reforms and bureaucracy, the Nordic model, civil society, democracy and the welfare state. He has been visiting scholar/professor at Stanford University, Harvard University and the University of California at Berkeley as well as EHESS, Paris, Max Planck, Cologne and Freie Universität Berlin. He is a member of several advisory and editorial boards and has been manager of several large-scale research projects. He is co-editor of the book series Nordic Studies in a Global Context at Routledge Press and *The Making and Circulation of Nordic Models* (Routledge 2022) as well as *Associative Governance in Scandinavia.*

Johan Strang is a Professor at the Centre for Nordic Studies, University of Helsinki. Trained as a philosopher, he has matured into an intellectual and political historian of the Nordic region. He has specialised on the relation between democracy and constitutional/human rights in the welfare state, as well as on Nordic cooperation and centre/periphery-relations in Europe. Recently, he has turned his focus towards developments at the end of the twentieth century, studying the transformation of Norden as a community and as a political argument since the end of the Cold War. Among his publications are *Nordic Cooperation: A European Region in Transition* (Routledge 2016), *Contesting Nordicness: From Scandinavianism to the Nordic Brand* (De Gruyter 2022, edited together with Jani Marjanen and Mary Hilson), *Nordic Peace in Question: A Region of and for Peace* (Routledge 2025, edited together with Christoper S. Browning and Marko Lehti) as well as the article "Rights against the welfare state" in the book *Nordic Neoliberalisms* (Routledge 2025, edited by Jenny Andersson and Chris Howell).

Part III
The Influences of Globalization and Liberalization

Chapter 5
How Financial Globalization Transformed the Nordic Model and Public Policy

Magnus Lindmark

Abstract This chapter examines the transformation of the Nordic welfare state during the 1990s and 2000s, focusing on the impact of financial market deregulation in the 1980s and 1990s. The Nordic Model, characterized by a sizable public sector, high living standards, and comprehensive social insurance systems, evolved under the influence of Social Democracy in the post-war era. Significant changes in this model were driven by structural shifts in the global economy, particularly the deregulation of financial markets and the advent of Information and Communication Technology (ICT). These changes facilitated the globalization of trade and finance, leading to increased acceptance of privately owned companies in welfare services and the liberalization of markets previously dominated by public monopolies. The chapter explores how these shifts influenced welfare policies, organizational structures, and economic dynamics, highlighting the interplay between global economic trends and local contexts. Through a detailed analysis of the Nordic countries' responses to these changes, the chapter sheds light on the complex interactions between economic forces and policy making that have shaped the modern Nordic welfare state then and now.

Keywords Welfare state · Nordic states · Regulatory reform · Financial markets · Globalization · Economic history

5.1 Introduction

During the 1970s, a distinct form of capitalism known as the Nordic Model evolved to varying degrees within Denmark, Sweden, Norway, Finland, and Iceland (Hilson, 2008). This model featured a market economy with strong elements of negotiation among political, trade union, and economic actors. Influenced by Social Democracy during the post-war period, the Nordic Model was characterized by a sizable public

M. Lindmark (✉)
Umeå University, Umeå, Sweden
e-mail: magnus.lindmark@umu.se

D. C. Nord (ed.), *Contemporary Nordic Public Policies*, Springer Polar Sciences,
https://doi.org/10.1007/978-3-032-11292-7_5

sector, high living standards, elevated taxes, and comprehensive social insurance systems.

However, during the 1990s and 2000s, the Nordic Model underwent significant changes characterized by liberal economic reforms. This chapter explores the main research questions: What were the primary drivers behind the transformation of the Nordic welfare state from the 1970s to the 2000s? And how did financial market deregulation and globalization influence this process? The objective is to analyze the transformation of the Nordic welfare state, focusing on the impact of financial market deregulation in the 1980s and 1990s, and to understand how structural shifts in the global economy, particularly the advent of Information and Communication Technology (ICT) and the globalization of trade and finance, contributed to the marketization of welfare services in the Nordic countries.

The Nordic welfare state featured generous and generalized social insurance systems and the free provision of welfare services. These services, including medical care, healthcare, schooling, and higher education, were offered either by the state, local governments, or municipalities. Other basic services such as telecommunications, postal services, electricity, radio, and television were almost exclusively provided by commercial enterprises on a monopoly market basis. From around 1990 to 2008, this Nordic Model underwent significant changes.

The main argument of this chapter is that the transformation of the Nordic Model was driven by a combination of economic crises, financial market deregulation, and the rise of new industries such as ICT. These factors facilitated the globalization of Nordic trade and finance, which led to an increased acceptance of privately owned companies providing welfare services and the liberalization of markets once previously dominated by public monopolies. While these general trends were similar across all the Nordic countries, different national variations emerged reflecting some of the distinctive economic, political, and social contexts within the region.

Following the example of Maddison (1991) the causal analysis that is set forth in this essay distinguishes between "ultimate" and "proximate" drivers. Ultimate causality involves consideration of historical processes, technical development and economic structural change (Schön, 2009). Proximate causality includes regulatory changes which affect economic initiatives. The interplay between these levels of drivers, it is argued, redistributed the political power of private and public interest groups, which became the direct proximate driver behind a reformed Nordic welfare state.

The basic idea advanced in this chapter is that the Nordic Model was developed during a post-war international economic regime under U.S. leadership that was characterized by strict regulations of financial markets. This regime began encountering severe economic difficulties in the 1970s. This was matched with declining productivity and a slowing of growth in sectors of the Nordic economy that had driven regional prosperity during the 1950s and 1960. Additionally, underneath the surface, a new emerging sector, Information and Communication Technology (ICT), needed necessary funding. This required capital, which would be made possible by the deregulation of capital markets in the 1980s. As argued in this chapter, the crisis faced by the post-war industries in the 1970s, and the rise of a new ICT industry, were ultimate drivers of the process that led to the transformation of the Nordic welfare state. The proximate driver was the deregulation of the financial markets and, more specifically, the remarkable growth in global and Nordic stock markets that followed.

5.2 Cracks in the Post-war Economic Order

The basic elements of the Nordic Model appeared in the Nordic countries during the interwar period of the twentieth century. However, the model further evolved during the region's period of strong economic growth during the 1950s and 1960s (This development is discussed in Chaps. 1 and 2 of this volume). The established postwar economic order, under American leadership, allowed for so-called "catch-up growth" by countries recovering from the ravages of World War II. The Nordic nations, with strong design capabilities and high levels of productivity, would soon "catch-up" with U.S. by engaging in trade and investing in American technologies.

Politically, the Cold War existed at the same time. While Denmark, Norway and Iceland joined NATO, Finland and Sweden remained unaligned. However, nearly all Nordic countries were integrated into the economies of the West under American leadership. The international monetary system, known as the Bretton Woods system, was a cornerstone of this integration allowing for the full convertibility of Nordic currencies at fixed exchange rates (Bordo & Eichengreen, 1993). The GATT, OECD, IMF and the World Bank were other mainstays of the system.

The Nordic Model evolved in an institutional setting where credit markets were heavily regulated. By controlling the money supply, domestic inflation could be monitored, which was a prerequisite for maintaining a fixed exchange rate under the Bretton Woods system. Likewise, control of the external flow of capital was a major pillar for post-war stabilization policies (Henrekson, 1988). These controls were more extensive in the Nordic region than in other Western European countries (Grytten & Hunnes, 2016, p. 230). Consequently, when the financial system was eventually deregulated in the 1980s, the effects were more significant there.

The Bretton Woods system was suspended in 1971, when the U.S. Federal Reserve could no longer defend the fixed dollar exchange rate to gold. After the collapse of the Bretton Woods system, the Nordics attempted to maintain fixed exchange rate regimes through various systems, such as the European 'currency snake' of 1972, as well as pegging their individual currencies to a "basket" of other international currencies. This turned out to be difficult undertaking within small, open economies such as those found in the Nordic region.

5.3 The Nordic Economies and the 1970s Crisis: Inner Diversity

On top of the financial uncertainties following the collapse of the Bretton Woods system, OPEC's 1973 oil embargo further exacerbated the emerging economic crisis for both the international and Nordic communities. The plummeting demand for oil tankers, the overproduction of industrial sectors in many Western countries, along with heavy government subsidization, all contributed to a serious global economic crisis by 1976. As profits fell in the Nordic region, several leading companies in heavy industries, especially in Sweden, required either extensive financial support from government or to be taken over by the state for reorganization (Schön, 2016). Government subsidies aimed at supporting such struggling industries caused significant national budget deficits. These deficits were financed through loans. As interest rates increased, the result was that interest payments became among the largest items for Swedish government expenditure. This further exacerbated the overall national budget deficit.

It is important to remember, however, that each of the Nordic countries was affected differently by the economic crisis of the 1970s. In comparison to the 1950s and 1960s, Danish and Swedish economic growth slowed dramatically by the mid-1970s. This was not just a response to the worsening international economic environment, but also the consequence of post-war "catch-up" growth with the U.S. being largely over. As their productivity gap with the U.S. narrowed, Danish and Swedish businesses were increasingly forced to adopt innovation-driven strategies. Therefore, the need for new investment in these firms was great, and as a consequence, their profitability fell.

The Finnish situation differed from that of Sweden and Denmark. In the mid-1970s, the country had lower income levels than other Nordic countries. The reason for this was a lower productivity rate, which also suggested that there were still good investment opportunities in Finland. Finnish "catch-up growth" would continue until the early 1990s (Hjerppe & Jalava, 2006). Finland was also able to reduce the effects of the oil crisis through favorable bilateral trade agreements with the Soviet Union (García-Iglesias & Kilponen, 2006). These also tended to counteract any downturns in the export-dependent Finnish engineering industry.

Norway had approximately the same income level as Sweden and Denmark in the mid-1970s. Reasonably, therefore, the Norwegian economy should have slowed down in similar ways. However, it did not, which was due in large part to the substantial oil discoveries made in the North Sea. However, the development of the Norwegian petroleum industry required large investments. This was borne by the international oil industry. The expanding oil sector meant that Norwegian economic growth could continue trend-wise in roughly the same way as it had during the 1950s and 1960s. Yet it was, however, far from certain whether the Norwegian "oil adventure" would eventually turn out to be successful. Iceland faced by far the highest inflation among the Nordics in the aftermath of OPEC Crisis of 1973. Also, at nearly the same time, the Anglo-Icelandic fisheries disputes erupted sowing additional seeds of uncertainty into the national economy of that country.

5.4 Deregulation of Global Financial Markets

The deregulation process of U.S financial markets had already commenced in the 1970s (Evanoff, 1985; Versluysen, 1988). To comprehend the significance of these events, we need to revisit the early years of the Carter administration. The beginnings of pension investment in private equity in the United States and Canada dates to the late 1970s (Fenn et al., 1997). Large institutional investors diversified into "non-traditional" asset classes such as private equity and real estate. By 1977, a shortage of capital available for financing new ventures became increasingly evident. Changes in regulations, taxes, and securities laws, influenced by the Employee Retirement Income Security Act (ERISA), aimed at revitalizing the venture capital industry. This was of significance as ongoing international structural change, a crisis in traditional industries, and an emerging ICT-sector, all required improved mobilization of investments.

Similar needs were also felt within the Nordic region, especially in Sweden, with its large manufacturing industry sector. Another important U.S. regulatory change taking place at this time was the liberalization of the Securities Exchange Act (SEA) of 1934 (Grullon & Michaely, 2002). As Straehl and Ibbotson (2017) have shown, the new 1982 legislation led firms to start repurchasing their stock *en masse*. Such actions contributed to a strong development and expansion of the U.S. stock markets.

The deregulation of financial markets in the Nordic region commenced in the early 1980s, starting with Denmark. The latter had removed all Bretton Woods style regulations from its credit markets in 1982 (Grytten & Hunnes, 2016). The Norwegian deregulation effort was a more gradual process characterized by incremental reforms (Berg & Eitrheim, 2009). Sweden and Finland followed suit approximately two years later (Jonung, 2003; 2008). By 1986, most regulations of capital markets in the Nordic region had been removed apart from the fixed currency exchange rate regimes which remained (Henrekson, 1988).

Also, within the Nordic region, the role of the stock exchange, itself, was transformed. Stock markets there experienced a massive inflow of capital due to relaxed taxation on household savings in stocks. Nordic stock markets were also affected by an inflow of foreign capital due to relaxed regulations. Finally, Nordic firms were engaging in strategic business operations which necessitated the buying and selling of stocks (Broberg, 2023; Øksendal et al., 2019). Despite the transformative nature of these reforms, there was only limited public debate in any of the Nordic countries regarding these changes.

The tangible effects of these reforms resulted in a housing boom in Denmark and Norway in the early 1980s. This was later mirrored in Finland and Sweden. The private sector capitalized on the growing asset prices, using them as collateral to absorb more debt. As bank lending, especially for property purchases, surged, the volume of bank loans also significantly increased. Consequently, the inflation rate escalated further. This resulted in interest rates, adjusted for inflation expectations, being either negative or close to zero which reinforced the private sector's demand for loans. The overall Nordic financial system experienced an era of extreme expansion, leading to problematic over-lending, particularly in the Finnish, Norwegian, and Swedish economies. Ultimately, this boom phase soon transitioned into a bust phase in the late 1980s and early 1990s.

The main components of the economic crisis of that period were inflation and fixed exchange rates. Inflation was driven by excess money supply due to deregulation. The continuing existence of fixed exchange rates within the Nordic countries meant that rising inflation did not lead to an automatic adjustment of the exchange rate. Therefore, devaluation pressures built up. Again, these tendencies were most evident in Sweden. As Swedish housing prices collapsed in 1990, and building societies and banks started to go bust at the same time, the confidence of international investors in Swedish and other Nordic banks declined. This led to increased speculation against the Swedish currency. In the autumn of 1992, the Swedish Central Bank chose to defend the exchange rate by introducing a staggering 500% marginal interest rate. This quickly led to a dead stop in the national economy, and ultimately, the end of the fixed exchange rate regime in the Nordic region. As banks needed to be bailed out, the public debt again increased. Unemployment rose sharply, and so did interest rates.

The other Nordic countries were also affected by the economic crisis, but Sweden and Finland were hit the hardest. These two countries responded to the crisis in different ways. Sweden bailed out banks and allowed the social security system to act as an "automatic stabilizer," which meant that social expenditures increased as unemployment rose, leading to even greater public debt. In contrast to Sweden, Finland maintained a tight fiscal policy, resulting in higher unemployment and harsher consequences for households.

The prevailing economic theory of the day was that high unemployment stemmed from high interest rates which often were a consequence of substantial public debt. This economic theory was backed-up with a strong narrative that suggested that international financial actors set the conditions for Nordic economic policy. Swedish Minister of Finance Göran Persson expressed his views on this with a biblical reference: "He who is in debt is not free," referring to his 1996 visit to New York financial markets. During this visit, Persson described the bankers he met as "sneering financial pups" ("hånleende finansvalpar"), highlighting the extremely challenging negotiations he faced in dealing with the Swedish public debt. In essence, the bankers were demanding additional public saving and a reduction in welfare spending. The existence of significant public debt necessitated a reformation of the welfare state.

Turning to Denmark, several austerity measures were taken in 1986 after a troublesome period of macro-economic imbalances that had existed since the 1970s (Abildgren & Thomsen, 2011). This became known as "The Potato Cure," referring to a strict expenditure diet. These deeply unpopular reforms included reducing public spending through cuts in government budgets and public services, increasing taxes to reduce the budget deficit, wage controls, and raising interest rates to limit inflation and stabilize the currency. The tax reform also imposed higher taxes on interest income and reduced tax deductions on interest payments. Thus, private consumption was suppressed. Likewise, private consumption was suppressed. As a result of the "The Potato Cure," the Danish inflation rate was on its way down just as Sweden and Finland were facing their most serious challenges. During the 1990s, Denmark experienced more of a period of zero growth, rather than a plummeting GDP as was seen in Finland and Sweden. The market reforms that followed were rolled out in a less panic-stricken manner in Denmark than in the other Nordic countries.

Norway also faced a severe economic downturn and a banking crisis during the late 1980s and early 1990s. It was, however, less severe when compared to Finland and Sweden. Nonetheless it led to a series of structural and institutional reforms within that country aimed at bolstering the economy. These included strengthening the regulatory framework for banks to ensure financial stability; monetary policy adjustments; a reduction of public spending; and new initiatives to enhance competitiveness and flexibility within the economy (Steigum & Thøgersen, 2015). At the same time, the Norwegians were also a bit lucky as rising oil revenues helped to mitigate their economic crisis.

5.5 Reforming the Nordic Welfare State

The economic crises of the 1990s necessitated significant reforms within each of the Nordic welfare states. They were driven by somewhat varying economic contexts and pressures as we have seen. Yet all of the subsequent national reforms were aimed at stabilizing public finances, reducing public debt, and increasing efficiency in welfare services (Andersen et al., 2017; Kananen, 2014).

By the mid-1990s, the economic situations among the Nordic countries differed to some degree which also resulted in some differences in specific responses. Sweden faced the highest national debt, amounting to 70% of GDP. This compared to 60% for Denmark and Finland, and only 40% for Norway. The recent collapse of the Soviet Union added additional pressure on Finland's economy. Denmark and Norway's efforts at gradual financial deregulation resulted in relatively stable banking systems. Both Sweden and Finland endeavored to right their own economic ships by successfully tapping into the ICT revolution with companies like Ericsson and Nokia leading the way.

A strong focus throughout the region was placed on increasing labor market participation. Reforms included tightening eligibility for unemployment and disability benefits, promoting active labor market policies, and encouraging work over welfare. For example, Sweden's "earned income tax reduction" reform of 2006 sought to increase the labor supply and reduce health insurance costs.

Pension systems were also restructured so as to ensure their long-term sustainability. This often involved shifting from defined benefit programs to defined contribution schemes, raising retirement ages, and linking benefits more closely to lifetime earnings. Norway's major reform effort was the establishment of an "Oil Fund" in 1990, later renamed "The Government Pension Fund Global" in 2006, which invested natural resource rents from oil extraction activities. Even though these oil revenues became extremely large, imitations were placed on how such oil fund money could be used to address domestic issues. It required strict political discipline. As Norwegian Prime Minister Jens Stoltenberg suggested, "We can't extract nurses from the North Sea," emphasizing the reality that Norway's oil wealth, alone, could not solve labor shortages or inflationary pressures in the domestic economy. Focused and restrained public spending would be required.

Globalization also imposed general economic pressures on each of the Nordic economies through increased competition. Combined with the need to reduce public debts and maintain international competitiveness, such pressures further underlined the need to reform the welfare state in order to successfully confront these economic realities. Cost-effectiveness in health and social services became a high priority, leading to measures such as decentralization, increased private sector involvement, and the introduction of market mechanisms within public services (See Chap. 6 of this volume).

Marketization and privatization also included the introduction of competition and private sector participation in public services such as healthcare, education, and elderly care (Burström, 2015). Public–Private Partnerships (PPPs) allowed private companies to supply public services while still being funded by the state. Sweden went the furthest in the direction of marketization and privatization with the introduction of "free schools" (friskolor) in 1992 and allowing joint stock companies to run publicly funded schools and healthcare institutions. Denmark increased private sector involvement in elderly care and social services. Finland's approach was more cautious, with gradual reforms aimed at increasing efficiency and cost-effectiveness within the existing system. Norway maintained a strong public sector, however, with limited private sector involvement compared to Sweden and Denmark. Iceland's small population and unique economic challenges also led to a more conservative approach to privatization.

Marketization and privatization were seen by many as necessary reforms so as to allow public saving by increasing efficiency and productivity in the welfare sector. The concurrent goal was to reduce public debt, a problem that had become difficult, over time, as such indebtedness was increasingly financed through the global financial markets. The remainder of this chapter examines the actual impacts of globalization on the Nordic countries in terms of trade and finance and their subsequent need for reform of the welfare state. The exceptional developments in the stock markets after 1980 are presented as a proximate driver behind these reforms.

5.6 Globalization of Trade as an Ultimate Factor of Change

If small economies, like those of the Nordics, are to thrive, they need to pursue an open trade policy so as to benefit from such exchange. Globalization of trade was ultimately driven during the last decade of the twentieth century by two basic innovations: ICT and the standardized container. The reduction in transportation and information costs that stemmed from these allowed for market expansion and the creation of global value chains. Political change, manifested in the collapse of the Soviet Union and the advent of the New Economic Policy of China also contributed to bring additional change to the existing system. All these were the proximate drivers behind trade as it became more globalized in the early 1990s. By comparing the development of globalized trade in the Nordic countries with other economies, we can get a sense of how these trends ultimately affected the reformation of the welfare state in the region.

Table 5.1 shows the KOF Swiss Economic Institute's Trade Globalization Index for the Nordics, the U.S., France, Spain and others states over several decades and allows for such comparison. Among the Nordics, Norway was the most commercially globalized country in the early 1970s, with Denmark a close second. The economies of Sweden, Finland and Iceland were less globalized. Overall, the Nordic were highly globalized countries already by the early 1970s.

However, what is interesting here is what happened the 1990s, when—with the exception of Norway—all of the Nordics experienced a sharp increase in the globalization of their trade An important factor behind this growth was the creation of the European Union (EU) and its internal market in 1993. The European Union included Denmark as a member from its inception. However, in 1995 Sweden and Finland also joined. Notably, Norway and Iceland secured access to the internal market through the Agreement on the European Economic Area (EEA) in 1994, but remained outside

Table 5.1 Overall trade globalization index for The Nordics, the U.S., and selected other countries, 1970–2020

	1970	1980	1990	2000	2010	2020
Sweden	56	60	63	78	80	79
Denmark	62	68	69	81	81	81
Norway	70	74	73	75	70	70
Finland	54	64	58	77	77	77
Iceland	52	54	61	75	72	72
USA	39	43	49	53	53	52
France	52	62	60	70	68	69
Spain	42	49	51	68	63	67
Canada	38	43	48	62	57	60
Japan	33	36	36	43	52	54

Source KOF Swiss Economic Institute (Dreher, 2006; Savina et al., 2019)

of the EU itself. However, as seen in Table 5.1, after the year 2000, the growth in trade globalization among the Nordics has been relatively small.

The surge in trade globalization across most of the Nordic countries during the 1990s was particularly pronounced in comparison to the trends observed in larger European nations, such as France and Spain, but also with regard to countries such as the U.S. and Japan. The pronounced economic diversity characterizing larger economies like the U.S., and to a lesser extent France and Spain, may have tempered their reliance on international trade during this period. But it was in the 1990s, that the Nordics were at the forefront of the globalization trend. It was during this decade that several Nordic companies successfully established global supply and value chains and witnessed them expand into the international marketplace. NOKIA, IKEA and Volvo were examples of such participants. While trade globalization presented both challenges and opportunities for Nordics, it was primarily the globalization of finance during the same period that was to have the greater effect on the transformation of the welfare state.

5.7 Financial Globalization as an Ultimate Factor of Change

The growth in financial globalization among the Nordic countries followed a more uniform pattern than trade as can be seen in Table 5.2. In the early 1970s, Sweden, Denmark and Norway were at roughly the same level of financial globalization, while Finland, and especially Iceland, were at lower levels. When compared to the position of the U.S., a general observation can be made that the Nordic countries were more highly globalized in trade, but less so in terms of financial globalization.

Table 5.2 Overall financial globalization index for the Nordics, U.S., and selected other countries (1970–2020)

	1970	1980	1990	2000	2010	2020
Sweden	41	50	70	87	88	88
Denmark	43	55	75	88	83	86
Norway	43	53	63	79	80	82
Finland	34	48	65	86	83	88
Iceland	26	25	33	62	62	61
USA	39	43	48	53	53	52
France	52	61	60	69	68	69
Spain	28	39	49	81	79	82
Canada	53	57	62	80	77	82
Japan	30	39	52	59	62	78

Source KOF Swiss Economic Institute (Dreher, 2006; Savina et al., 2019)

From 1975 onward, however, a new trend toward additional financial globalization began in all the Nordic countries with the exception of Iceland. This process lasted until the early 2000s. Sweden, Denmark and Finland were then at approximately the same high level, while Norway's financial globalization was 10% lower. For Iceland, the increase financial globalization was concentrated in the 1990s and then plateaued and remained lower than other Nordic countries.

For the Nordic countries, the rate of growth in financial globalization was larger than what was seen in trade globalization. Also, the growth in financial globalization were more uniform across the Nordics as compared to trade, thus reflecting less Nordic inner diversity. Even more interestingly, financial globalization started earlier in the Nordic countries than in the other national economies shown in Table 5.2. Thus, the growth of financial globalization in the Nordic countries *preceded* the reformation of the welfare state and was in position to have constituted an ultimate factor of change with regard to it.

The traditional notion that large public debts drove up interest rates, which, in turn, caused high unemployment has been mentioned. The standard recommended cure was to reduce government spending and to amortize the public debt. In principle, this could have been done in the Nordic states during the period of the economic crisis of the 1990s. The "cheese slicer" principle could have been adopted without significant organizational changes to the welfare state. Yet this approach was not taken. Instead, organizational changes, not the least marketization, were substantial. The reason for doing so requires a more specific explanation than simply the broad impact of financial globalization. The role which stock markets played in the reform of the Nordic welfare state needs to be highlighted.

5.8 Stock Markets as a Proximate Factor of Change

Figure 5.1 compares the U.S. Dow-Jones Index with the Stockholm Stock Exchange General Index (OMX) over time and normalized by GDP. A rising trend line means that stock markets are growing faster than GDP. The most striking feature of the graph is the changing slope in the early 1980s, indicating that the stock market indices started to grow at a considerable faster rate in relation to GDP greater than had been seen previously. The U.S. stock market had grown faster than its GDP in the 1950s, and then had transitioned to a growth rate lower than its GDP from the mid-1960 onwards. Figure 5.1 also demonstrates that Sweden exhibited a similar pattern, with a robust stock market, growing slightly faster than GDP from the early 1950s until the mid-1960s. This decline from the mid-1960s onward fits well with Robert Gordon's (2016) argument that the established technologies of the day gradually lost their power to generate economic growth.

In Swedish case, the OMX index increased by a factor of 10 during the 1980s, and again, starting in the early 1980s, and continuing until the early 2000s, the Dow Jones and OMX indices rose sharply relative to GDP. In Sweden, the OMX index increased by a factor of 10 during the period and in the U.S. by a factor of 5. This exceptionally

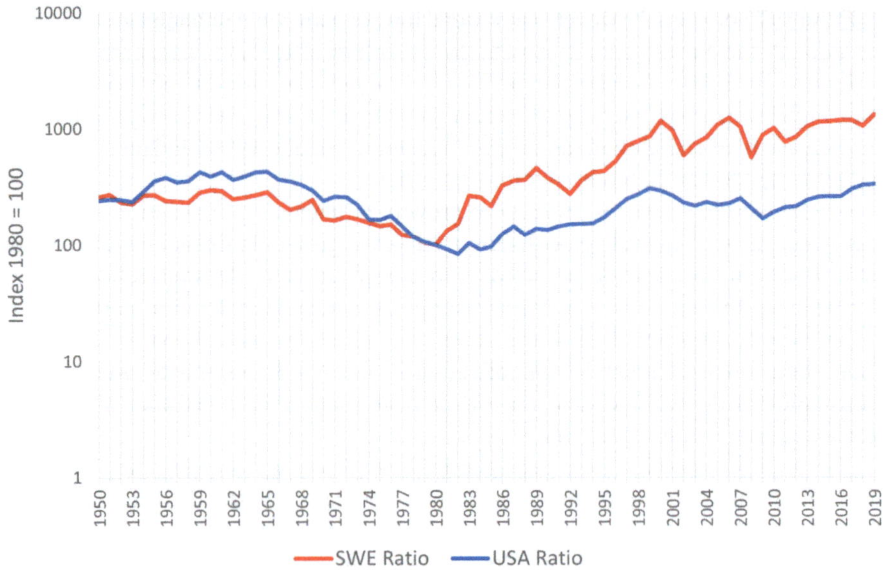

Fig. 5.1 Stock index-to-GDP ratio for Sweden and the USA (1950–2020), log scale on *Y*-axis. *Sources* own estimates based on: Waldenström (2014); Feenstra et al (2015); MacroTrends. Dow Jones—DJIA —100 year historical chart

high growth in Sweden can be attributed to the devaluation of the Swedish Krona in 1982 and the increased in U.S. interest rates during 1979, which appreciated the U.S. Dollar. All of this gave a profound boost to the Swedish export industries such as SKF. Likewise, the ICT sector, representing a new wave of general-purpose technologies, boosted companies like Ericson. Pharmaceutical producers, such as Astra, were other winners. Firms like IKEA, Volvo and Electrolux successfully created global supply chains.

Added to this wave of good economic results was the deregulations of financial markets. This increased demand on the stock markets in both in Sweden and the U.S. One should also note that Sweden introduced tax reforms in the late 1970s and early 1980s, which encouraged household saving in equity funds through tax deduction. This entire period of rapid development of the stock market lasted from the early 1980s to the early 2000s. A booming stock market was in evidence in all Nordic countries, though strongest in Sweden and Finland during the critical 1980s. The Dot Com Crisis, the internet financial bubble of the early 2000s, marked the end of this period.

Figure 5.2 illustrates the performance of stocks in all of the Nordic countries, except Iceland. It shows that all countries experienced a stable stock index-to-GDP ratio until the mid-1960s, followed by a decline until the early 1980s (with Norway declining earlier and more continuously). Thereafter, all countries saw a rapid increase in the ratio until the early 2000s, with more modest growth afterward.

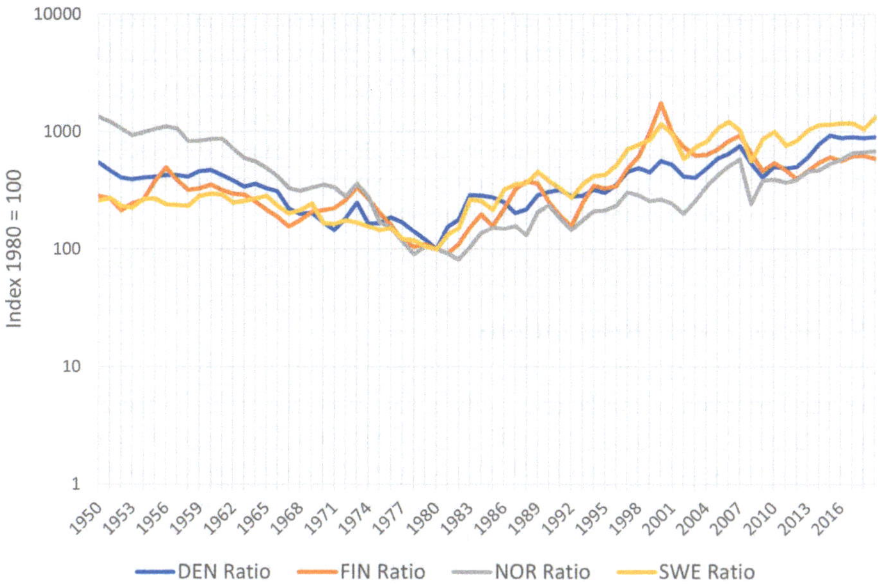

Fig. 5.2 Stock index-to-GDP ratio for Denmark, Finland, Sweden, and Norway (1950–2020), log scale on *Y*-axis. *Sources* own estimates based on Feenstra et al (2015); Waldenström (2014); Statistikbanken (Statbank.dk); Soumen Pankki; Eitrheim (2004)

Notably, Norway continued to experience growth beyond 2000, while Finland saw a decline between 2002 and 2012 in the aftermath of the dot.com crisis.

Based on this historical review of Nordic stock market performances, we can draw the following conclusions. From 1950 to the mid-1960s, stock markets grew at approximately the same rate as GDPs. From the mid-1960s to the early 1980s, the ratio declined in all countries, with Norway experiencing an earlier and more pronounced decline. What is interesting in this period, from a public policy perspective, is that if one wanted to finance a hospital, it would matter little economically whether you issued stocks and tapped into the financial pool of the Stockholm stock exchange, or if you levied taxes, tapping into GDP. The political pressure for marketization of the public sector was indeed minimal.

However, from the early 1980s onwards, it became much better to issue stocks. This meant that the booming stock markets of the period provided a mechanism for the marketization of the welfare state within the Nordic region. Nordic stock companies could raise capital more easily and cheaply than governments. This created pressure to open up state regulated monopoly markets for private investment. This included the reorganization of public monopolies into stock-based companies.

As deregulated markets offered opportunities to earn interest on capital, Nordic business and financial interests found a strong motive to lobby for such policy initiatives. Also, since Nordic stock companies could potentially provide welfare services at lower costs, taxes could be reduced. Likewise, the public debt could be reduced, without the hardships of traditional austerity. This became an attractive option for most Nordic politicians, including conservatives and right-wing Social Democrats (kanslihushögern). Finally, it was argued the middle-class households could benefit from lower taxes through such measures while enjoying similar levels of welfare. All of this built political support for such welfare policy reforms. From such a perspective, it is evident how developments within the Nordic financial markets contributed to the transformation of the Nordic welfare state.

5.9 Some Final Observations

The transformation of the Nordic welfare state from the 1970s to the 2000s was driven by a combination of economic crises, financial market deregulation, and the rise of new industries such as Information and Communication Technology (ICT). These factors facilitated the globalization of trade and finance, leading to increased acceptance of privately owned companies in welfare services and the liberalization of markets previously dominated by public monopolies. Overall, the transformation of the Nordic welfare state could be characterized by periods of intense reform activity, particularly during the 1980s and 1990s, followed by more gradual public policy adjustments and refinements in subsequent decades. This can be summarized as follows:

1980s: This decade saw the beginning of significant changes, particularly in response to the economic crises of the 1970s. The focus was on stabilizing public finances and reducing public debt, which led to initial steps towards marketization and deregulation. Extensive deregulation of financial markets marked this period, setting the stage for subsequent transformations.

1990s: The pace of reforms accelerated during the 1990s, marked by a new era of economic policy in the wake of accelerating globalization. Reforms included the privatization of state-owned enterprises and the introduction of market mechanisms within public services. The economic recession of the early 1990s that was felt in many Nordic countries further pushed their governments to implement efficiency-driven reforms.

2000s: The early 2000s continued the trend of market-oriented reforms, with a strong emphasis on improving efficiency and reducing public expenditure. This period saw the expansion of private sector involvement in welfare services, including healthcare and education, through mechanisms like competitive tendering.

2010s and beyond: The pace of reform has slowed somewhat. The focus of public policy has shifted towards fine-tuning the existing market-based systems and addressing issues of equity and access. There has also been a growing debate about the balance between market efficiency and the universal principles of the welfare state.

The main argument of this chapter has been that the transformation of the Nordic welfare state was driven, in part, by a combination of economic crises, financial market deregulation, and the rise of new industries such as ICT. These factors facilitated the globalization of trade and finance, which, in turn, led to increased acceptance of privately owned companies providing welfare services along with the liberalization of markets previously dominated by public monopolies.

Each Nordic country's unique economic, political, and social contexts, especially in the early 1990s, shaped their particular path of reform. For instance, Sweden and Denmark embraced market-oriented responses more extensively, introducing private providers and competitive mechanisms in sectors like healthcare and education. In contrast, Norway and Finland were more cautious, maintaining a stronger role for the public sector and implementing market reforms more selectively.

Also, the economic crises of the 1970s and 1980s affected each Nordic country differently, influencing their urgency and approach to reforms. Countries with higher public debt and economic instability, like Sweden, were more aggressive in implementing market-based solutions to reduce public expenditure. Additionally, the strength and orientation of the political parties within each country played a crucial role. In Sweden, the Social Democrats, despite their traditional support for the welfare state, led significant market reforms in response to economic pressures (See Chap. 4 of this volume). In contrast, Norway's political landscape, with a stronger presence of parties favoring public sector solutions, resulted in more cautious reforms.

Finally, the strength and influence of financial interest groups varied across the Nordic countries. In Sweden, strong financial sectors and active lobbying by business groups facilitated the adoption of market-oriented reforms. In contrast, the other Nordic countries with less influential financial sectors, like Finland, saw slower and more limited marketization. All of these factors suggest a pattern of some inner diversity within the region.

A crucial factor in this whole transformation of the Nordic welfare state was the role played by stock markets. The deregulation of financial markets led to a significant increase in stock market activity, providing a mechanism to mobilize cheap capital for investments. The booming global stock markets created a formidable mechanism for the marketization of the Nordic welfare state, as stock companies could mobilize capital more cheaply than other types of organizations. After the early 2000s, the stock index-to-GDP ratio remained relatively stable, with modest fluctuations, suggesting that while stock markets continued to play a role, their transformative pressure on welfare reform diminished. This pressure led to the opening of regulated monopoly markets private firms and the reorganization of public monopolies into stock companies. These developments within the financial markets, therefore, played a pivotal role in the transformation of the Nordic welfare state. After the early 2000s, the stock index-to-GDP ratio remained relatively stable, with modest fluctuations, suggesting that while stock markets continued to play a role, their transformative pressure on welfare reform diminished.

In conclusion, it is suggested that the transformation of the Nordic welfare state was a complex process influenced by global economic trends, technological advancements, and country-specific factors. The marketization of welfare services was driven by the need to reduce public debt and improve efficiency and was facilitated by financial market deregulation and the rise of new industries. Further research is needed to explore the varying degrees of marketization across different Nordic countries and the role of financial interest groups in shaping welfare state reforms.

References

Abildgren, K., & Thomsen, J. (2011). A tale of two Danish banking crises. *Monetary Review, 50*(1), 121–142.

Andersen, J. G., Schøyen, M. A., & Hvinden, B. (2017). Changing Scandinavian welfare states: Which way forward? In P. Taylor-Gooby, B. Leruth, & H. Chung (Eds.), *After austerity: Welfare state transformation in Europe after the great recession* (pp. 89–114). Oxford University Press.

Berg, S. A., & Eitrheim, Ø. (2009). Bank regulation and bank crisis: The main developments in the Norwegian regulatory system before, during and after the banking crisis of 1988–92 (Working Paper No. 2009/18). Norges Bank.

Bordo, M. D., & Eichengreen, B. (1993). *A retrospective on the Bretton Woods system: Lessons for international monetary reform.* University of Chicago Press.

Broberg, O. (2023). Vi skapar en ny marknad! Vill du vara med? Historien om när börsen blev ekonomins naturliga mittpunkt. In J. Andersson, N. Glover, O. Husz, & D. Larsson Heidenblad (Eds.), *Marknadens tid* (pp. 267–288). Nordic Academic Press.

Burström, B. (2015). Sweden—Recent changes in welfare state arrangements. *International Journal of Health Services, 45*(1), 87–104.

Eitrheim, Ø., Gerdrup, K., & Klovland, J. T. (2004). Historical stock price indices in Norway 1814–2003 (Occasional Papers No. 35). Norges Bank.

Evanoff, D. D. (1985). Financial industry deregulation in the 1980s. *Economic Perspectives, 9*(5), September/October, 3–5.

Fenn, G. W., Liang, N., & Prowse, S. (1997). The private equity market: An overview. *Financial Markets, Institutions and Instruments, 6*(4), 1–106.

García-Iglesias, C., & Kilponen, J. (2006). Monetary aspects of a changing economy. In J. Ojala, J. Eloranta, & J. Jalava (Eds.), *The road to prosperity: An economic history of Finland* (pp. 187–216). Suomalaisen Kirjallisuuden Seura.

Gordon, R. J. (2016). *The rise and fall of American growth: The U.S. standard of living since the civil war.* Princeton University Press.

Grullon, G., & Michaely, R. (2002). Dividends, share repurchases, and the substitution hypothesis. *Journal of Finance, 57*(4), 1649–1684.

Grytten, O. H., & Hunnes, A. (2016). *Krakk og kriser i historisk perspektiv.* Cappelen Damm Akademisk.

Henrekson, M. (1988). Räntebildningen i Sverige. SNS Förlag.

Hilson, M. (2008). *The Nordic model: Scandinavia since 1945.* Reaktion Books.

Hjerppe, R., & Jalava, J. (2006). Economic growth and structural change: A century and a half of catching-up. In J. Ojala, J. Eloranta, & J. Jalava (Eds.), *The road to prosperity: An economic history of Finland* (pp. 33–64). Suomalaisen Kirjallisuuden Seura.

Jonung, L. (2003). Den finansiella revolutionen, 90-talskrisen och inflationsmålet. In L. Jonung (Ed.), *På jakt efter ett nytt ankare: Från fast kronkurs till inflationsmål* (pp. 25–50). SNS Förlag.

Jonung, L. (2008). Lessons from financial liberalisation in Scandinavia. *Comparative Economic Studies, 50*(4), 564–598.

Kananen, J. (2014). *The Nordic welfare state in three eras: From emancipation to discipline.* Routledge.

Maddison, A. (1991). *Dynamic forces in capitalist development: A long-run comparative view.* Oxford University Press.

Schön, L. (2009). Technological waves and economic growth—Sweden in an international perspective 1850–2005 (CIRCLE Electronic Working Paper Series No. 6). CIRCLE, Lund University.

Schön, L. (2016). *An economic history of modern Sweden.* Routledge.

Steigum, E., & Thøgersen, Ø. (2015). A crisis not wasted—Institutional and structural reforms behind Norway's strong macroeconomic performance. In T. M. Andersen, M. Bergman, & S. E. H. Jensen (Eds.), *Reform capacity and macroeconomic performance in the Nordic countries* (pp. 246–273). Oxford University Press.

Straehl, P. U., & Ibbotson, R. G. (2017). The long-run drivers of stock returns: Total payouts and the real economy. *Financial Analysts Journal, 73*(3), 32–52.

Versluysen, E. L. (1988). Financial deregulation and the globalization of capital markets. Background paper for the 1988 world development report. World Bank.

Øksendal, L. F., Brautaset, C., & Ecklund, G. J. (2019). Børsen. Markedsplass og møteplass 1819–2019. Universitetsforlaget.

Data Sources

MacroTrends. Dow Jones—DJIA—100 year historical chart. https://www.macrotrends.net

Statistikbanken (Statbank.dk). Tabell MPK13. Statistics Denmark. https://www.statbank.dk

Suomen Pankki (Bank of Finland). Finnish share indices. https://www.suomenpankki.fi

Dreher, A. (2006). Does globalization affect growth? Evidence from a new index of globalization. *Applied Economics, 38*(10), 1091–1110.

Eitrheim, Ø., Gerdrup, K., & Klovland, J. T. (2004). Historical stock price indices in Norway 1814–2003 (Occasional Papers No. 35). Norges Bank.

Feenstra, R. C., Inklaar, R., & Timmer, M. P. (2015). The next generation of the Penn world table. *American Economic Review, 105*(10), 3150–3182.

Savina, G., Haelg, F., Potrafke, N., & Sturm, J.-E. (2019). The KOF globalisation index—Revisited. *Review of International Organizations, 14*(3), 543–574.

Waldenström, D. (2014). Swedish stock and bond returns, 1856–2012. In R. Edvinsson, T. Jacobson, & D. Waldenström (Eds.), *Historical monetary and financial statistics for Sweden, Volume II: House prices, stock returns, national accounts, and the riksbank balance sheet, 1620–2012* (pp. 223–292). Sveriges Riksbank and Ekerlids Förlag.

Magnus Lindmark is professor in Economic History at Umeå University, Sweden. His research explores how economies grow and change over time. His research primarily focuses on economic growth, natural resources, and environmental issues, as well as financial history. His preferred analytical approach involves examining structural changes as outcomes of the convergence of distinct historical development paths—each rooted in different economic and political contexts—at specific points in time. Guided by economic theory, his work aims to understand institutional transformations at the systemic or structural level.

Chapter 6
The Diffusion of Trust Reforms Within the Nordic Countries

Tina Øllgaard Bentzen and Marte Winsvold

Abstract The Nordic countries are renowned for their remarkable levels of trust, a quality that has received increased examination over recent years in conjunction with the adoption of New Public Management practices within the public sector. Trust has emerged as a pivotal focal point for consideration and evaluation of these Scandinavian efforts. National trust-reforms have been implemented in Denmark, Sweden and Norway over recent decades with the explicit goals of reducing bureaucracy, promoting decentralization, and revitalizing societal confidence. Concurrently, the need for societal trust has permeated policymaking, and taken a central position on the agenda of most Scandinavian municipalities. This chapter provides a comprehensive exploration of the diffusion of such trust-reforms and their significant role in shaping the current landscape of public sector organizations. Drawing from diverse Scandinavian settings, the chapter examines both the similarities and differences found in these trust-reforms undertakings across the region. It concludes with a critical examination of the prospects, challenges and inherent paradoxes associated with the pursuit of trust through public sector reforms and the implications of such an undertaking for future public policymaking.

Keywords Public sector reform · Trust · Magic concepts · Frontline employees · Administrative burdens · Leadership

6.1 Introduction

Compared to almost all other countries in the world, trust in fellow citizens, public institutions, and democracy is a central core of the Nordic welfare societies (Edelman inc., 2022; Svendsen & Svendsen, 2016). Trust is often referred to as the 'Nordic

---

T. Ø. Bentzen (✉)
Roskilde University, Roskilde, Denmark
e-mail: tinaob@ruc.dk

M. Winsvold
Institute for Social Research, Oslo, Norway

D. C. Nord (ed.), *Contemporary Nordic Public Policies*, Springer Polar Sciences,
https://doi.org/10.1007/978-3-032-11292-7_6

gold'—a fundamental value that permeates the fabric of the Nordic countries, playing a crucial role in facilitating interactions in all areas of society, including the workplace (Andreasson, 2017; Svendsen & Svendsen, 2015).

This chapter explores the phenomena of the Nordic Trust Reforms, which for more than a decade have played a central role in policy and practice of Nordic governance. We offer an overview of the Danish, Swedish and Norwegian trust reforms and discuss similarities and internal variations among them. The chapter also engages in a critical discussion of the problems and prospects of using trust as a magic concept in public sector reforms. We argue that Nordic trust reforms provide both inspiration and invaluable knowledge to all policymakers seeking to build a more trustworthy and responsive public sector.

6.2 Why Trust Reforms in a High-Trust Corner of the World?

The high levels of trust found in the Nordic countries are rooted in their histories of relative peace and stability, prosperity, and economic equality (Bergh & Bjørnskov, 2011) and interlinked with the Nordic welfare system, known for its extensive public services and high levels of taxation (Knudsen & Rothstein, 1994). Scandinavian citizens enjoy free healthcare, education, social support, and other benefits that are highly dependent on generalized trust that ensures tax compliance. Although occasional public scandals may spur debate, residents of the Nordic countries generally trust their institutions to provide fair governance and they expect compliance by their fellow citizens (Torfing & Bentzen, 2022). High levels of trust are also sustained by traditions of political collaboration across political parties and the co-creation of public policy with other societal actors (Torfing et al., 2019; Windvold & Bentzen, 2020).

Scandinavian work culture is generally characterised by extensive delegation, minimal power distance and an informal relationship between managers and employees. Trust in the workplace is partly seen as a consequence of a strong cooperative tradition between managers and employees and close collaboration with trade union organizations (Byrkjeflot, 2024; Skirbekk & Grimen, 2012). Leadership practises rely heavily on democratic values and a tradition of active involvement, emphasising a softer approach to power over an authoritarian approach (Frost & Moussavi, 2011).

There are, however, signs today that trust within the Nordic public sector has been challenged by increasing societal turbulence, demographic developments and the influence of control-oriented public sector reforms. In particular, the New Public Management (NPM) reforms introduced in the 1980s have been criticized for straining traditionally high levels of trust, particularly between public sector leaders and their employees (Bentzen, 2019a; Torfing & Bentzen, 2022; Vallentin & Thygesen, 2017). Increasing challenges with recruiting and retainment of skilled public professionals have only accelerated this debate about rebuilding trust with the public employees, contributing to a political and administrative pushback against NPM, most notably in the form of so-called "Trust Reforms". Labors unions have raised widespread concerns among public sector employees being overly controlled and spending excessive time complying with detailed documentation requirements. Growing bureaucratic burdens and performance control are experienced as lack of trust in employees (LO & OAO, 2012).

Rooted in this critique, the Nordic Trust Reforms seek to promote goals of reducing dysfunctional control, decentralising decision-making and strengthening trust with frontline employees in the public sector by granting them more autonomy (Bentzen & Bringselius, 2023; Siverbo et al., 2023; Vallentin & Thygesen, 2017). Not only have such reform efforts been launched in Denmark, Sweden and Norway to promote trust as a core agenda in the public sector at the national level (Johnsen & Svare, 2024), but "trust" has also become an organizational beacon today in most Scandinavian municipalities (Siverbo et al., 2023).

While reform efforts to rebuild citizens' trust in political leaders or institutions are common in European reforms efforts (OECD, 2017; Van de Walle, 2013), the Nordic Trust Reforms are exceptional in their distinctive focus on building trust with front-line employees in the public sector. Although trust-based governance has also been discussed in Finland, the emphasis on and proliferation of these trust reforms at both national and municipal levels are most prominent in Denmark, Sweden, and Norway (Siverbo et al., 2024). Likewise, while Iceland has engaged in trust-focused reforms, these efforts have been primarily concentrated on restoring citizens' trust in political institutions (Sigurgeirsdóttir & Johnsen, 2018). As such, we see that the "trust reform" effort also reflects some degree of inner diversity among the Nordic states.

Since problems of waning trust is a general challenge encountered today in most Western democracies, some scholars argue that governance-inspired reforms, such as the Nordic Trust Reforms, are crucial for ensuring that the public sector remains attractive to a skilled workforce (Krogh & Triantafillou, 2024; Nyhan, 2000; Torfing, 2023). Hence, the Nordic Trust Reforms may offer valuable examples and insights for other countries facing challenges in retaining competent employees as a foundation for a robust public sector.

The Trust Reforms in Scandinavia have tended to be motivated by criticism of previous governance ideals and practices (Torfing & Bentzen, 2022; Torfing et al., 2020). In public administration, it is widely recognized that three central paradigms have significantly influenced the shaping and development of the public sector in the region over time. They are: (1) Old Public Administration (OPA), which emerged in the late 19th and early twentieth centuries, emphasizing hierarchy, rules, and the virtues of equality and transparency in public services; (2) New Public Management (NPM), gaining prominence from the 1980s, which introduced market mechanisms, competition, and a focus on public sector efficiency and performance; and (3) New Public Governance (NPG), emerging in the early 2000s, which advocates for strengthening networks with other community sectors and promoting the co-creation of policy with citizens and other societal actors to innovatively address societal challenges (Torfing et al., 2020).

The Nordic Trust Reforms often embody a critique of the excessive rules and time-consuming hierarchical decisions of Old Public Administration, which are seen as barriers to flexibility and innovative problem-solving for public employees. Even more evident is the critique of New Public Management reforms, which promoted market thinking and performance-orientated management (Hood, 1991; Torfing & Bentzen, 2022). A widespread concern regarding the decades of public sector reforms influenced by the ideology of New Public Management is that they have led to escalating control measures, significant transaction costs and frustration among the front-line professionals within the public sector (Van De Walle, 2013). While NPM essentially promised to reduce bureaucracy and simplify the public sector, critics argue that the reality in many public sector organisations turned out to be quite different. New layers of administration and additional documentation requirements have strained time for core tasks, innovation and, above all, eroded employee motivation (Hood & Dixon, 2015; Torfing & Bentzen, 2020).

Even though NPM reforms have generally been implemented more moderately in the public sector of the Nordic countries than in other Western European countries, neoliberal management practises there have been strongly criticised for increasing mistrust of employees and limiting their professional autonomy (Almqvist et al., 2013; Greve & Ejersbo, 2003; Klausen, 1998). In contrast to this, recent initiatives in trust reforms within Scandinavia appear to advocate approaches to leadership and governance in the public sector that promote participation, reduce power imbalances, and distribute decision-making authority (Bentzen, 2018; Johnsen et al., 2022; Siverbo et al., 2023).

The emphasis on trust in these reform efforts often resonates with New Public Governance ideals, including concepts such as social capital, co-creation, interactive governance, and collaborative networks. All of these emphasise the importance of building relationships and partnerships (Kramer, 2003; Six & Sorge, 2008; Svendsen & Svendsen, 2016; Torfing et al., 2019; Vallentin & Thygesen, 2017). Hence, while the Scandinavian trust reforms are primarily aimed at strengthening or restoring the trust of public sector employees (Krogh & Triantafillou, 2024), some scholars argue that they are also an integral part of New Public Governance ideals of engaging and mobilising citizens and civil society actors (Bentzen, 2018, 2021; Van De Walle, 2013). Recently, some other scholars have argued that, in practice, the Nordic Trust Reforms may actually be less about replacing existing paradigms than pragmatically reconfiguring existing paradigms into slightly new hybrids (Johnsen & Svare, 2024; Torfing & Bentzen, 2022).

In addition to addressing the perceived deficiencies of previous management approaches, the Trust Reforms within the Nordic countries have come in response to some basic challenges regarding the recruitment and retention of employees within the public sector. With an aging population and competitive employment opportunities in the private sector, it has become progressively more difficult to secure and maintain a skilled workforce in public services within the Nordic region. This is particularly true when skilled personnel feel that their views and input are not being encouraged and when they are being faced with excessive administrative burdens. Not only is this corner of the world known for its high level of trust, but also for substantial levels of happiness and contentment. The latter are being increasingly challenged today in the public sector by growing numbers of workers who express growing job dissatisfaction related, in part, to a deficiency of trust.

6.3 A History of the National Nordic Trust Reforms

As shown in Fig. 6.1, the first of the national trust reforms was introduced in Denmark in 2013. It was followed by similar efforts in Sweden and later ones in Norway. The figure also illustrates that the original trust reform in Denmark had a relatively short lifespan before being replaced by new undertakings coming from a new government. Although trust as a policy headline was soon abandoned, it remained a permeating policy concept and objective of all subsequent Danish governments. Hence, we refer to all of these undertakings, collectively, as the Danish Trust Reforms.

6.3.1 The Danish Trust Reform(S)

Growing criticism of the problematic side-effects of NPM shaped the debate about the decreasing levels of trust within the Danish public sector for several years (Gjørup et al., 2007). However, the first formal mention of the *Danish Trust Reform* only emerged in 2011. It was outlined in the Foundation Paper of the newly elected government that was comprised of the Social Democratic Party, the Danish Social

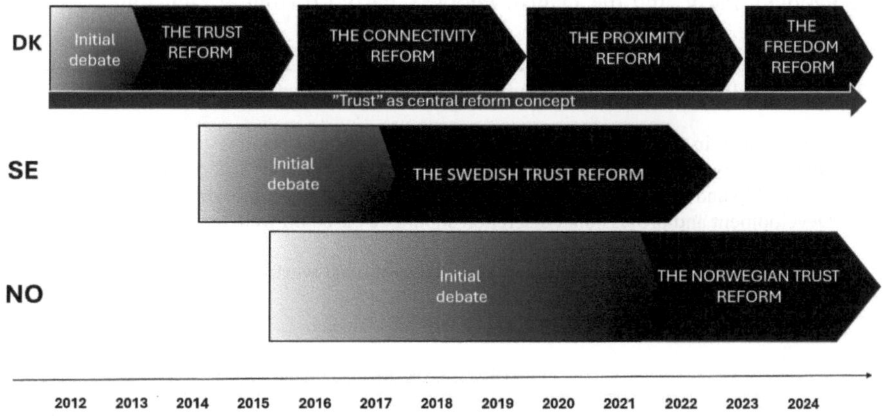

Fig. 6.1 The Danish trust reforms

Liberty Party, and the Socialistic Peoples Party. In part, it announced that: "The government aims to launch a reform centred on trust, leadership, professionalism, and de-bureaucratization" (Regeringen, 2011).

The Danish Government's announcement highlighted the pressing need to de-bureaucratize the public sector, addressing years of accumulating bureaucracy, rules, and control. It ambitiously stated that oversight not directly contributing to citizens' value should be relinquished, allowing space for trust to flourish. Despite extensive general debates in the Danish press regarding the necessity of a "trust reform", it took almost two more years before its specific features were formally introduced in a policy document outlining "seven principles for modernizing the public sector" (See Fig. 6.2). Included among these principles were ambitious goals of founding leadership and management in the public sector on trust and openness. It also called for responsibility, clear goals, and results.

This concise formulation, was a hallmark of Denmark's cooperative tradition, formed as it was in collaboration with unions, municipal representatives (Kommunernes Landsforening), and the Danish regions. Also, beyond outlining basic principles, it left for the public organizations, themselves, to discuss and implement the details of the endeavour. As a consequence, very few specific policy initiatives were launched under this initial trust reform effort in Denmark. Some governance laboratories were established and a new Centre of Innovation was created. However, it remained somewhat unclear how these were connected to the reform's aspirations (Bentzen & Bringselius, 2023). Yet the Danish unions actively supported these trust reforms by crafting comprehensive outlines and recommendations for embedding trust principles in all public organizations. While the Danish Trust Reform garnered vocal support among unions and municipalities, it increasingly faced criticism among the general public regarding its vagueness, inadequate funding, and perceived lack of progress (Allentoft, 2014).

In 2015, this first Trust Reform effort was concluded with the election of a new coalition government comprised of the Left Danish Liberal Party, Conservative People's Party, and the Danish People's Party (Torfing & Bentzen, 2022). As

The Seven Principles of the First Danish Trust Reform Effort

1. Management in the public sector should focus on goals and results.
2. Dialogue, openness, and clear objectives should be the basis for task completion.
3. Leadership and management should be founded on trust and responsibility.
4. Development and professional discretion should be based on well-founded documentation.
5. Task completion should be based on knowledge of what works.
6. Leadership should promote innovation.
7. Public service should engage the resources of the citizens.

(Regeringen, Akademikerne, Danske Regioner, FTF, 2013)

Fig. 6.2 The seven principles of the Danish trust reforms

a result, the initial Danish Trust Reform officially ended only a few years after it begun. However, its goals for expanding trust within the public sector were not abandoned. Instead, they were continued in a new "***Connectivity Reform***" launched by the new coalition government. This policy initiative underscored existing critiques of the New Public Management and the continuing need to rebuild trust within the public sector. The new government suggested that: "We aim to eliminate meaningless rules, formulas, and rigid control. New Public Management has become excessive. Providing welfare requires trust in public employees" (Finansministeriet, 2017).

Thus, while "connectivity" became a new watchword for reform, trust remained a core concept and ambition within its undertakings. Unlike the previous trust reform efforts, where only limited policy documents and initiatives were produced, this new Danish Connectivity Reform presented numerous programs that outlined its ambitions and initiatives (Finansministeriet, 2017). Besides introducing four guiding policy beacons, the reform also involved an initiative aimed at simplifying laws governing Danish job centres and an array of "freedom experiments" at the institutional and municipal levels. These experiments were later evaluated by the National Research and Analysis Centre for Welfare, which generally found that such bureaucratic loosening created a positive impact on task quality and employee motivation (Hjelmar, 2021). Additionally, a Commission of Leadership, convened researchers and leaders from both public and private organizations to offer recommendations for the further development of leadership within the public sector. Sophie Løhde, the Minister of Internal Affairs, received relatively positive reviews of the resulting policies that were formed in agreement with opposition parties and involved relevant stakeholders.

This proved to be fortuitous since in 2019, the Social Democratic Party's election victory led to the end of the specific Connectivity Reform. It was replaced by its own "*Proximity Reform*," that continued to prioritize the goal of trust. The new Danish prime minister, Mette Frederiksen, emphasized trust 19 times in her 2019 opening speech to parliament, consistently highlighting the need to empower public employees and to "set free" the public sector (Regeringen, 2018).

The new Proximity Reform notably introduced extensive "liberation" experiments in municipal areas such as day-care, elderly care, and schools. In 2020, seven Danish municipalities were granted increased autonomy to act within so-called "Welfare Agreements," allowing them freedom from a larger array of state regulations. After receiving positive feedback from these policy decisions, the remaining 95 Danish municipalities were offered similar opportunities in 2022 (Hjelmar et al., 2022). Evaluations once again pointed to positive effects of the free municipality experiments. Extended freedom to act resulted in improved solutions for vulnerable citizens through streamlined services, a significant reduction in response times for child welfare cases, and increased citizen satisfaction (Hjelmar et al., 2022). However, an election once again changed the course of Danish trust reforms.

In October 2022, an unprecedented government consisting of the Social Democratic Party, the Liberal Party of Denmark, and the newly formed Moderate Party gained power. Despite their political diversity, their common ambitions to build trust in society and the public sector allowed the Danish reform effort to continue.,

However, the title of the reform effort changed with the new government. "Freedom" became the new policy headline. As in previous programmes, "trust" and opposition to burdensome control within the public sector remained as core ambitions. The new Danish government argued that:

> Decades of management regimes and a growing bureaucracy have eroded time for relationships between people. (…) A nursing home should once again be more of a home than an institution, and we have fundamental trust that parents, children, teachers, and leaders are the ones who create the best schools. (Regeringen, 2022)

While the established focus on trust in reshaping the public sector persisted, under the new "*Freedom Reform*," ambitions for de-bureaucratization were not only aimed at liberating the public employees, themselves, but also the citizens (Regeringen, 2022). While on the one hand distancing itself from the New Public Management, the new Danish government also embraced classical neoliberal ideas such as "free choice" and privatization. Interestingly, such neo-liberal ideas were also combined with other governance ideals calling for stronger collaboration with civil society, individuals, and the private sector (Regeringen, 2024). Programme initiatives aimed at reducing administrative burdens, establishing a simplified and more collaborative model for supervision within Eldercare and reducing, uncoordinated supervision from diverse authorities. Inspired by the successful "welfare agreements" of the previous government, an extensive de-regulation of schools was also launched in 2024.

Summing up, the original version of Danish Trust Reform was relatively vague and short-lived. However, it was followed up by a line of other public sector reforms with a strong emphasis on trust as a core value in governance. While initially introduced by a Social Democratic government, "trust" as a concept in reform of the public sector became almost apolitical in the sense that it was adopted by all of the major parties in Denmark. This evolution is summarized in Table 6.1.

As we shall see, the idea of using trust as a beacon in public sector reform also spreads across the border to Sweden.

6.3.2 *The Swedish Trust Reform*

In Sweden, the trust reform effort[1] arose amidst critical debates about issues stemming from NPM-oriented control. Inspired in part by the Danish Trust Reform, the Swedish Social Democratic government-initiated discussions about trust as a central governance ambition during its 2014 election campaign. In 2016, the Minister of Public Administration Ardalan Shekarabi officially announced a broadscale Swedish trust reform. He noted: "There is a necessity to develop trust-based governance of welfare services, founded on mutual trust among stakeholders. This applies to both

[1] For the sake of comparison, we refer to the trust initiative in Sweden as a reform. However, it can be debated whether "reform" is the appropriate term, or if it should rather be called a "trust movement."

Table 6.1 The Danish trust reforms

The Danish Trust reforms				
Reform:	The trust reform (Tillidsreformen)	The connectivity reform (Sammenhængsreformen)	The proximity reform (Nærhedsreformen)	The freedom reform (Frisættelsesreformen)
Period of time	2011–2015	2017–2019	2019–2023	2023–?
Political foundation	The Social Democratic Party The Danish Social Liberal Party The left Green (Socialistic Peoples Party)	Left Danish Liberal Party Conservative Peoples Party The Danish Peoples Party	The Social Democratic Party	The Social Democratic Party The Liberal Party of Denmark The Moderate Party
Ambassador of reform	Minister of Internal Affairs, Margrethe Vestager (The Danish Social Liberal Party)	Minister of Public Innovation, Sophie Løhde (Left Danish liberal party)	Prime Minister, Mette Frederiksen (The Social Democratic Party)	Prime Minister, Mette Frederiksen (The Social Democratic Party)
Administrative anchoring	Ministry of Internal Affairs	Ministry of Finance	Ministry of Finance (Political Centre of Welfare)	Ministry of State (Secretary of Liberation)

state governance of municipalities and county councils, as well as their management of institutions" (Regeringen, 2016).

A central aspect of the Swedish trust reform was the establishment of a Trust Commission ("The Delegation of Trust-Based Governance") that was rooted in academia. From 2016 to 2019, the Commission developed theoretical frameworks, articulated examples of trust-based leadership, and made recommendations. It disseminated these through seminars, conferences, and workshops (Bringselius, 2017). While the Swedish unions played a minimal role in these efforts, academic research served as a driver for the initiative, deploying a wide array of resources and research projects on trust-based management and employee empowerment (Bentzen & Bringselius, 2023). The Swedish Trust Commission also formulated seven principles (see Fig. 6.3), with trust as the foremost, designed to inspire and guide local implementations of trust-based leadership (see e.g., Tillitsdelegationen (2019)). All state agencies were encouraged to adopt these and introduce new programs for state employees that embodied them.

The objective of the Swedish Trust Commission was to develop practices aimed at achieving a more trust-based system of governance within agencies and creating a more coordinated public sector. Several pilot projects involving municipalities, regions, and universities were launched to test such trust-based governance initiatives. The Trust Commission also organized a series of networks and forums where researchers, leaders, and employees could discuss relevant topics. Pilots such as "Tillitsverkstaden" allowed six state agencies to address specific identified challenges by implementing trust-based management (Bentzen & Bringselius, 2023). The Commission's remit also included proposing methods for sharing experiences and best practices in conceptual and organizational development across central government agencies. This would ensure their continuation even after the commission's work was completed.

Seven Principles of the Swedish Trust Reform

1. **Trust:** Strive to choose to trust those you collaborate with as a starting point and have positive expectations.
2. **Citizen Focus:** Put the citizen and user experience and knowledge in focus and try to understand what is valued.
3. **Holistic View:** Encourage everyone in the management chain to actively and collectively take responsibility for the whole, i.e., co-leadership, and collaborate across boundaries.
4. **Scope for Action:** Delegate authority and mandate, welcome co-determination.
5. **Support:** Ensure close-to-operations, service-oriented, and qualified professional, administrative, and psychosocial support in the core business.
6. **Knowledge:** Reward knowledge development, continuous learning, and practice based on science and proven experience.
7. **Openness:** Strive for openness by sharing information, welcoming dissent, and respecting criticism.

(SOU, 2019)

Fig. 6.3 Principles of the Swedish trust reforms

Table 6.2 The Swedish trust reform

Reform	The Swedish trust reform
Period of time	2016–2021
Political foundation	Social democrats
Ambassador of reform	Minister of Public Administration, Ardalan Shekarabi
Administrative anchoring	Head of the Trust Commission, Laura Hartman (2016–2019) and later Christina Forsberg (2018–2019)

Despite numerous research reports and the production of extensive material, the Swedish Trust Reform effort faced criticism for being too vague about how it should be implemented in practice within the different levels of the public sector (Nosrati et al., 2022). Some proposed reform elements were also criticized for simply making adjustments to policy rather than bringing about radical change (Nagell, 2021). Similarly, the undertaking was criticized for being primarily focusing on leadership styles and neglecting crucial aspects of public steering, regulation, and control (Siverbo, 2023).

Overall, the Swedish Trust Reform could be characterized as a thorough, research-based approach leading to an abundance of documentation, reports and recommendations. However, the actual extent of implementation into practice has been debated. Since 2019, no new initiatives have emerged from the Swedish Government, and in 2020, the Trust Commission was officially abandoned. While some claim that this was a clear sign of a waning interest in Swedish trust reform (Aspøy, 2023; Siverbo, 2023), others suggest that trust reform remains a priority within the country and that politicians generally maintain a loyalty to trust-based management principles (Bringselius, 2023). Table 6.2, summarizes the main features of the Swedish Trust Reform initiative.

6.3.3 The Norwegian Trust Reform

As was the case in Denmark and Sweden, the introduction of trust reforms in Norway was triggered by frustration with the New Public Management (NPM). Discussions in the Norwegian public media since 2013 revealed concerns with excessive and overly strict controls within the public sector stemming from neoliberal approaches to government. The discussion gained momentum in 2015 and the idea of a trust reform initiative gradually developed. The left of the political spectrum, led by the Norwegian Labour Party, harnessed the growing dissatisfaction with NPM approaches and promoted the concept of trust-based reforms. In the run-up to the 2021 election campaign, the Labour Party included "trust reform in the public sector" within its party manifesto and made it a key campaign issue. Following the Labour Party's coalition victory with the Agrarian Party, the implementation of nationwide trust reform became a cornerstone of their joint political agenda. The Government noted that: "The trust reform of the public sector aims to restore trust, time, and empowerment to those at the forefront and core of the welfare state, such as teachers, nurses, police officers, and general practitioners" (Regeringen, 2021).

This political statement provided guiding principles for the reform. It emphasized the need for expansion of professional autonomy among public employees, increased employee involvement, and a reduction of goals and oversight by authorities. The reform effort was presented in an ongoing narrative from which we have extracted seven key principles that epitomise the essence of Norwegian national trust reform (see Fig. 6.4).

The precise framework of the Norwegian reform effort was intended to be developed "in collaboration with those affected" (Regeringen, 2021). The Ministry of Local Government and Regional Development was entrusted with the coordination of the trust reform. Together with the Norwegian Agency for Public and Financial Management, they organised a series of conferences to initiate discussions on the framework and the possible content of the trust reform. The government ministries and their subordinate agencies were instructed to study, develop, and implement measures that were in line with the principles outlined above. In addition, efforts were made to gather knowledge and the broader experience of other Scandinavian trust reform efforts Direktoratet for økonomistyring (2023). This initiative reflected an interest in policy learning from within the Nordic region. As had been the case in Denmark, the trade unions of the country also actively participated in these trust reform discussions and helped to shape the contours of the undertaking (Nagell, 2021). The important influence exerted the Norwegian trade unions was underlined in both the repeated recommendation that trust reform should be based on the Norwegian labour market model and in the emphasis given to securing the agreement of trade unions in major public policy reforms.

While the concept of trust reform in Norway was met with broad approval from both trade unions and public sector stakeholders, there was also some criticism that the reform effort lacked focus and a specific plan action (Aspøy, 2023) The various possible interpretations and expectations associated with the rather abstract concept

7 Principles of the Norwegian Trust Reform

1. Offer frontline employees more professional freedom
2. Reduce unnecessary reporting and documentation requirements
3. Give local and regional authorities more room for manoeuvre
4. Strengthen co-determination and participation opportunities for employee organisations
5. Improve management, organization and coordination
6. Increase the use of trust-based leadership
7. Reduce the reporting and documentation obligations of the supervisory and control authorities in a targeted manner

Fig. 6.4 Principles of the Norwegian trust reforms

Table 6.3 The Norwegian trust reform

Reform	The Norwegian trust reform
Period of time	2021–
Political foundation	The Labour party and the Agrarian party
Ambassador of reform	Leader of Labour Party, Jonas Gahr Støre
Administrative anchoring	Ministry of Local Government and Regional Development and Ministry of Digitalisation and Public Governance

of "trust" became apparent. Furthermore, some of these debates highlighted certain dilemmas related to the scope of trust and raised the question of whether granting greater discretion to employees could potentially exacerbate inequality (Bentzen, 2022; Nosrati et al., 2022).

In the autumn of 2023, the national government invited Norwegian municipalities and counties to apply to participate in trial programs of trust reform within the public sector. This initiative offered the selected municipalities an opportunity to explore innovative approaches to task management and the granting of exemptions from existing regulations as had been done previously in Denmark's "freedom" experiments. Municipalities and counties were invited to propose experiments covering all public sectors, with a particular priority given to the integration of social services and the provision of health care, and education. Climate adaptation and land use policies, including compliance with the Norwegian Planning and Building Act, were also included. It is noteworthy that 87 municipalities (around one fifth of all Norwegian municipalities) have expressed an interest in participating in the trial program, which is scheduled to start in 2025.

The primary reform goal of the Norwegian trust initiative has been to restore the viability of public sector employees. Its focus has been on securing professional autonomy, encouraging greater employee involvement, and reducing excessive bureaucratic control. The main features of the Norwegian Trust Reform are summarized in Table 6.3.

6.4 Policy Learning

The somewhat sequential development of the Nordic trust reforms, raises the question of policy-learning across initiatives and countries. To what an extent did national policymakers explore previous experiences with trust reforms in their policy-design and implementation? There is little doubt that the Swedish policy-makers found inspiration in the Danish trust-reform effort and actively sought knowledge about reform initiatives there and their results (Aspøy, 2016). Norwegian policymakers appeared even more eager to build on previous trust-reforms within the Nordic area and systematically gathered information about the experiences of both national and municipal trust reforms in Denmark and Sweden (DFØ, 2023; Nagell, 2021).

Numerous Norwegian delegations spent time with both Swedish and Danish municipalities and researchers to harvest insights for their own trust reform effort. An interesting occurrence, perhaps reflecting cross-country inspiration, was the development of a seven-part list of principles for the trust-reform in both Denmark and Sweden. Interestingly, while the Norwegian government did not replicate this list directly, emphasising the importance of local adaptation, Norway's second-largest city, Bergen, adopted its own version of the seven principles in their local trust reforms (bergen.kommune.no, 2024). As we shall discuss later, trust reforms seem to have been adopted and shaped with some recognition of both the commonalities and differences among the Scandinavian countries. As such, they demonstrated both a certain degree of exceptionalism and inner diversity in their undertakings.

6.5 Across the Nordic Trust Reforms: Similarities and Variations?

Looking at the Nordic countries, it can be observed that trust has spread and persisted as a central headline for public sector reforms for more than a decade. Several common features can be identified in the Nordic trust reforms. They share the core objectives of decentralising authority and decision-making to local levels of government; empowering frontline staff; and reducing administrative burdens to allow more flexibility in public service delivery and decision-making. In this regard, they are distinctive or "exceptional" when compared to other types of public sector reform efforts found elsewhere in the world (Aucoin, 1990; Greve et al., 2020).

However, there is also some degree of inner diversity among the Nordic trust reforms within the public sector. One significant difference is the kind of political support these reforms have received. In contrast to Sweden and Norway, where trust reforms have been explicitly linked to classical socialist politics, Danish trust reforms have evolved as an almost apolitical activity. Denmark's experience with its trust reforms over the past decade is notable for their cross-party acceptance. Each successive Danish government, regardless of its political colour, has moved forward with trust reforms that, despite changing headlines, have been only slightly modified from those of its predecessor (Bentzen, 2020; Bentzen & Bringselius, 2023). This has not been the case in Norway or Sweden where national reforms have carried the clear imprint of specific political party endorsements. At the local level, however, the trust reforms have tended to be seen as politically neutral in all of the Scandinavian countries (Siverbo et al., 2023).

Additionally, the reform efforts of the public sector in each country have evolved in distinct ways. Denmark stands out as the originator and earliest of the reformers (Bentzen, 2015; Torfing & Bentzen, 2022). The issue of trust has remained on the national agenda for some time. Denmark's specific undertakings, however, have been relatively short-lived in nature due to repeated changes in government. Earlier initiatives have been given new labels as successive governments have come to power.

Despite largely consistent content that promotes trust-building with frontline public employees, Danish trust reforms have been constantly revised and "re-wrapped" with new labels. This practise has been criticised for being superficial and insufficient over time in that it does not allow for real, long-term change to mature through reform at a national level. Yet at the same time, trust as an agenda item and as an organizational practise has spread rather convincingly within Danish municipalities (Aspøy, 2023; Bentzen, 2022; Bentzen, 2021; Johnsen et al., 2022; Siverbo et al., 2023).

The Swedish Trust initiative, in contrast, was well-researched and thoroughly argued from the outset as a consistent package. However, its inability to secure formal implementation and earn continued priority by the Swedish government raises questions as to its overall effectiveness. As for the Norwegian Trust reform effort, it is still too early to predict either its public policy trajectory or its eventual outcome.

As previously noted, all the Nordic trust reforms have tended to define themselves either in opposition or as an alternative to neoliberal ideals and practices found in New Public Management. Criticism of excessive control and bureaucratization of the public sector has been a common theme found in all of the trust building initiatives within the Nordic countries (Siverbo et al., 2023). However, there are also real differences between them if you take a closer look at the specific national views of the NPM. While the trust reform effort in Norway has actively opposed any form of privatisation and outsourcing in the public sector, these elements of the NPM are addressed more pragmatically in the Denmark where there is not the same clear distancing (Aspøy, 2023; Bentzen, 2022). For example, the recent 'Freedom Reform', launched by the current Danish coalition government, presents NPM themes such as free choice for citizens and privatisation alongside their new efforts to build trust in and reduce control over public employees (Regeringen, 2023).

One possible explanation for these differing reform approaches, could be the contrasting economic contexts found in the two countries. Denmark has faced considerable economic pressure over the last three decades, compelling policymakers to continuously cut costs, optimise processes and experiment with market dynamics in the public sector (Bentzen et al., 2019; Svendsen & Svendsen, 2015). Conversely, Norway, with its substantial oil reserves, has experienced less financial pressures and therefore has had less incentive to seek market-based solutions and budget cuts. However, recent Norwegian studies of trust-based leadership within the Oslo Municipality suggest that some Norwegian trust reforms are less ideologically pure in practice than their policy pronouncements may suggest. Despite their call for major reforms of the current system, most of these initiatives typically involve adjustment and supplementation of existing governance forms rather than radical replacement (Johnsen & Svare, 2024).

While it is important to consider what the trust reforms are moving away from; it is equally important to understand what these reforms are moving towards. Although securing greater trust, autonomy and influence for employees in the public sector is a common goal for all the Nordics, there are also differences in how this goal is to be achieved among the three countries. The Norwegian initiative has repeatedly emphasised the importance of basing trust reforms on strengthened co-operation between participants, safeguarding employees' rights, and strengthening the professional competence of managers. Today in Norway, the latter are now expected to lead in a more democratic way. In many respects, the contours of the Norwegian trust reform effort has been aimed at returning to a pre-NPM public sector—that is, to what can be termed "professional governance" (Torfing et al., 2020).

Denmark also favours greater professional freedom and cooperation with the trade unions. However, the recent Danish trust reforms have also emphasised the relationship of trust with civil society and citizens. Therefore, the current Danish Trust Reform promotes the mobilising of resources among citizens, users, and other civil society actors in a more collaborative form of welfare and policy development. Such an approach requires not only new expertise in the public sector, but also employees who can collaborate closely with users, relatives, and other relevant interest groups (Sørensen & Bentzen, 2019; Torfing et al., 2020). While such New Public Governance-inspired ideas increasingly dominate the Danish trust reform effort, they are less pronounced in Norway and Sweden. One possible explanation for this difference could be that Danish municipalities, due to an extensive amalgamation reform in 2007, are notably bigger than their Swedish and especially Norwegian counterparts. Consequently, establishing contact and dialogue with citizens can be more challenging in a Danish context, which could make new governance approaches in these areas more relevant (Bentzen et al., 2019; Houlberg & Ejersbo, 2020).

6.6 The Nordic Trust Reforms: Problems and Prospects of a Magic Concept

Considering the recent history of trust reforms in Sweden, Denmark and Norway, it is fair to say that trust has become a central concept public sector reforms within the region. The separate Scandinavian trust reforms reveal a common fascination with trust as a governance agenda not only at the national but also at municipal levels (Siverbo, 2023; Bentzen & Winsvold, 2025), where substantial activity regarding trust reforms can be observed both in Denmark (Bentzen 2018; Vallentin & Thygesen, 2017), Sweden (Elmersjö & Sundin, 2021; Håkansson, 2022) and Norway (Bentzen et al., 2024; Johnsen & Svare, 2024). However, it is also evident that the impact and value of trust reforms have also sparked debate. Critics have pointed to issues such definitional vagueness, lack of funding, and insufficient initiatives as real barriers to achieving the desired goals of trust (Aspøy, 2023; Bentzen & Bringselius, 2023).

Assessing the long-term prospects for these national trust reforms is a major challenge, as the aspirations and objectives of these reform efforts are often formulated in relatively abstract terms. Some academics argue that the term 'trust' itself contributes to this ambiguity and vagueness, as it possesses the characteristics of so-called 'magic concepts.' (Bentzen et al., 2024; Carey et al., 2018; Pollitt & Hupe, 2011). As in other fields, organizational ideas for the public sector typically follow a bell-shaped life cycle, where they are adopted, embraced, and eventually discarded over time (Abrahamson & Fairchild, 1999; Boxenbaum & Jonsson, 2008). These conventional organizational concepts are hence anticipated to fall out of favour and diminish in influence over time. Alternatively, owing to their immediate applicability, attractiveness, widespread acceptance, and ease of promotion, "magic concepts" are more likely to persist on an organizational agenda for extended periods of time (Boswell, 2023). "Trust" has all the qualities of being a "magic concept" that comes with pronounced both strengths and weaknesses (Bentzen, 2019b). In avoiding specific ideological or interest groups ties, "trust" enjoys broad appeal and allows the idea to cover extensive domains, often with multiple, ambiguous, and sometimes conflicting definitions (Pollitt & Hupe, 2011). The substantial rhetorical value of magic concepts such as trust and their captivating nature render them appealing to public policymakers and leaders who utilize these concepts to promote, validate, and garner support for particular causes or agendas (Boswell, 2023). Magic concepts are excellent remedies for spurring engagement, mobilizing support, and setting new directions, but they are also fuzzy, hard to conceptualize and tend to suppress conflict by appearing univocally good or positive. These qualities seen in trust as a magic concept allows for diversity in translation but also challenge concrete implementation. While "trust" is easy to "sell" because of its omnipotent, positive vibe, it is far from simple to translate into policy as it tends to resist being pinned down (Bentzen et al., 2024).

Simultaneously, the same "magic" abilities of trust could explain why the Nordic trust reforms, despite continuous criticism, appear to spread, diffuse and stick on the policy agendas of these countries (Siverbo et al., 2023). Pollitt and Hupe, in their seminal work (2011), warned against the "seductive power" of magic concepts in public sector reforms, which they feared would lead to public resources being spent with little effect. This concern is shared among scholars who brand such popular, organizational concepts as superficial trends that primarily lead to symbolic alterations, resource wastage and unnecessary complexity (Abrahamson & Fairchild, 1999; Boxenbaum & Jonsson, 2008). Yet other authors stress that these metaphorical concepts can also act as catalysts for new perspectives and necessary organizational innovation. Boswell (2023), for example, argues that organizational actors may embrace magic concepts to stimulate necessary discussions and change, providing both procedural and substantive value. Thus, its invocation may not lead directly to substantially more trust in the system, but it may help confront and recalibrate detrimental organizational practices as well as catalyze new ones (Bentzen et al., 2024).

6.7 Conclusion

These Nordic efforts underscore the importance of trust as a central concept in public sector reform. Against a backdrop of exceptionally high levels of societal trust, these various initiatives aim to counteract the continuing negative effects of New Public Management practices of the past by seeking to decentralize decision-making, empower frontline employees, and promote a more flexible, less bureaucratic, public sector. However, while such trust reforms may be used to solve certain inherited public sector dysfunctions in the Nordic states, they can also frequently rely upon variants of both classical bureaucratic and neoliberal management practices and governance approaches (Johnsen & Svare, 2024).

Generally, however, these trust reforms reflect a shift towards more democratic and participatory models of governance that emphasize the professional autonomy of public servants and the reduction of excessive control and documentation (Bentzen, 2022; Bringselius, 2017). Despite these noble goals, trust reforms within the Nordic countries have also sparked considerable debate and criticism (Allentoft, 2014; Aspøy, 2023; Johnsen et al., 2022). The term "trust" itself, with its broad appeal and unclear definition, can been described as a "magic concept" that simultaneously encourages commitment and at the same time challenges concrete implementation difficult.

The Danish experience with trust reform in the public sector could be characterized by frequent changes in reform labels and a pragmatic approach to trust and market dynamics. This contrasts significantly with more ideologically consistent but less proactive Swedish and Norwegian approaches. These differences reflect the influence of varying national economic constraints and political contexts on the development and implementation of trust reforms. They also illustrate the overall contending pulls of commonality and inner diversity found within Nordic public policy today.

The Nordic Trust initiatives, despite their diversity and sometimes lack of focus, constitute distinctive type of public sector reform which has played a significant role in welfare organizations in this high-trust corner of the world. In their ability to attract skilled personal and reduce administrative burdens, the Nordic experiences with trust-reforms could be invaluable reference for policy makers and government leaders in other countries seeking to strengthen their own public sector. While some countries may focus narrowly on rebuilding citizens' trust in public institutions, the Nordic countries highlight the prospects of starting from within by building trust with front-line employees.

References

Abrahamson, E., & Fairchild, G. (1999). Management fashion: Lifecycles, triggers, and collective learning processes. *Administrative Science Quarterly, 44*(4), 708–740.

Allentoft, N. (2014, June 18). Tillidsreform i tomgang: Stigende frustrationer over offentlig modernisering. *Denoffentlige.dk.*

Almqvist, R., Catasús, B., & Wällstedt, N. (2013). New Public Management har skapat rädda organisationer. *Dagens Nyheter*.

Andreasson, U. (2017). *Trust—The Nordic gold*. Nordic Council of Ministers.

Aspøy, A. (2016). Tillitsreform i Skandinavia. *Stat & Styring, 26*(3), 14–16.

Aspøy, A. (2023). Hvor mye tillit kan vi ha til at regjeringen virkelig tar tillitsreformen seriøst? *Aftenposten.*

Aucoin, P. (1990). Administrative reforms in public management: Paradigms, principles, paradoxes and pendulums. Governance, 3(2), 115–137. https://doi.org/10.1111/j.1468-0491.1990.tb00111.x

Bergh, A., & Bjørnskov, C. (2011). Historical trust levels predict the current size of the welfare state. *Kyklos, 64*(1), 1–19.

Bentzen, T. Ø. (2015). *Tillidsbaseret styring og ledelse i offentlige organisationer—I springet fra ambition til praksis*. Roskilde University.

Bentzen, T. Ø. (2018). Tillidsbaseret styring og ledelse i offentlige organisationer (1. udg.). Djøf / Jurist- ogØkonomforbundet.

Bentzen, T. Ø. (2019a). The birdcage is open but will the bird fly? The interplay between institutional and interactional trust in public organizations. *Journal of Trust Research, 9*(2), 185–202.

Bentzen, T. Ø. (2019b). The translational diamond: Robust translation of magic concepts in public organizations. *International Journal of Public Leadership, 15*(2), 94–112.

Bentzen, T. Ø. (2020). Samskabt styring: Nye veje til afbureaukratisering. Samfundslitteratur.

Bentzen, T. Ø. (2021). Samskabt styring: Potentialer og udfordringer. Politica – Tidsskrift for Politisk Videnskab, 53(3), 261–279. https://doi.org/10.7146/politica.v53i3.130456

Bentzen, T. Ø. (2022). Veje og vildveje set fra et dansk perspektiv. *Stat & Styring*.

Bentzen, T. Ø., & Bringselius, L. (2023). Translating "trust" robustly in public sector reforms: A comparative study of the Danish and Swedish Trust Reform. *International Public Management Journal*.

Bentzen, T. Ø., Lo, C., & Winsvold, M. (2019). Strengthening local political leadership through institutional design: How and why. *Local Government Studies, 46*(3), 483–504.

Bentzen, T. Ø., Siverbo, S., & Winsvold, M. (2024). Keeping the magic alive: The multiple functions of magic concepts. *Public Administration Review*.

Bentzen, T. Ø., & Winsvold, M. (2025). Unveiling the double-edged sword of magic concepts in public sector reforms: Why capacity to cope with paradoxes makes all the difference. Perspectives on Public Management and Governance. Advance online publication. https://doi.org/10.1093/ppmgov/gvaf016

Bergen kommune. (2024). Tillitsreformen. https://www.bergen.kommune.no/tillitsreformen Direktoratet for økonomistyring. (2023). Tillitsarbeid i praksis. Statens arbeidsgiverportal. https://www.arbeidsgiver.dfo.no/tillitsarbeid-i-praksis

Boswell, J. (2023). *Magical thinking in public policy: Why naive ideals about better policymaking persist in cynical times*. Oxford University Press.

Boxenbaum, E., & Jonsson, S. (2008). Isomorphism, diffusion and decoupling. In R. Greenwood, C. Oliver, R. Suddaby, & K. Sahlin-Andersson (Eds.), *The Sage handbook of organizational Institutionalism* (pp. 77–101). Sage.

Bringselius, L. (2017). *Tillitsbaserad styrning och ledning—Ett ramverk*.

Bringselius, L. (2023). Trust-based public management: Conceptualization and lessons from the Swedish trust reform. In C. Greve, & T. Ysa (Eds.), *Handbook on strategic public management*. Edward Elgar.

Byrkjeflot, H. (2024). A Nordic model of associative governance? In *Associative governance in Scandinavia* (pp. 23–56). Routledge.

Carey, G., Malbon, E., Carey, N., & Newman, E. (2018). Strange magic: What can the emergence of "magic concepts" tell us about policy implementation? *Policy Design and Practice, 1*(3), 169–182.

DFØ. (2023). *Lærdommer fra arbeid med tillitsreform i skandinaviske land* (DFØ-notat 2023:1). Direktoratet for forvaltning og økonomistyring.

Direktoratet for økonomistyring. (2023). Tillitsarbeid i praksis. Statens arbeidsgiverportal. https://www.arbeidsgiver.dfo.no/tillitsarbeid-i-praksis

Edelman inc. (2022). *Edelman trust barometer 2022*.

Elmersjö, M., & Sundin, E. (2021). Intra-organisational trust and home care services: A study of the process of implementing trust-based practices in municipal eldercare in Sweden. *Scandinavian Journal of Public Administration, 25*(2), 3–21.

Finansministeriet. (2017). *Sammenhængsreformen. Borgeren først—en mere sammenhængende offentlig sektor*. Finansministeriet.
Frost, M., & Moussavi, F. (2011). *Trust in the Scandinavian workplace*. Sage Publications.
Gjørup, J., Hjortdal, H., Jensen, T., & Lerborg, L. (2007). Tilgiv os—vi vidste ikke hvad vi gjorde. *Politiken*.
Greve, C., & Ejersbo, N. (2003). When public-private partnerships fail: The extreme case of the NPM-inspired local government of Farum in Denmark. *Droit Public et Moderniser l'État, 3*(6).
Greve, C., Ejersbo, N., Lægreid, P., & Rykkja, L. H. (2020). Unpacking Nordic administrative reforms: Agile and adaptive governments. International Journal of Public Administration, 43(8), 697–710. https://doi.org/10.1080/01900692.2019.1645688
Håkansson, H. (2022). Contradictions of ordered trust: Trust-based work and conflicting logics in municipal care. Nordic Journal of Working Life Studies, 12(3), 3–22. https://doi.org/10.18291/njwls.130174
Hjelmar, U. (2021, January 14). Ni år med frikommuner har skabt bedre løsninger for borgerne. Berlingske. https://www.vive.dk/da/udgivelser/ni-aar-med-frikommuner-har-skabt-bedre-loesninger-for-borgerne-15820/
Hjelmar, U., et al. (2022). *Midtvejsevaluering af velfærdsaftaler*. VIVE.
Hood, C. (1991). A public management for all seasons? *Public Administration, 69*(1), 3–19.
Hood, C., & Dixon, R. (2015). *A government that worked better and cost less?* Oxford University Press.
Houlberg, K., & Ejersbo, N. (2020). Municipalities and regions—Approaching the limit of decentralization? In P. M. Christiansen, J. Elklit, & P. Nedergaard (Eds.), *The Oxford handbook of Danish politics* (pp. 141–159). Oxford University Press.
Johnsen, Å., Svare, H., & Wittrock, C. (2022). Tillitsbasert styring og ledelse i offentlig sektor i Norge? *Nordisk Administrativt Tidsskrift, 99*(1).
Johnsen, Å., & Svare, H. (2024). *Tillitsreformer og tillitsbasert styring og ledelse*. Fakbokforlaget.
Klausen, K. K. (1998). Indikatorer på NPM's gradvise men begrænsede gennemslag i Danmark. In K. K. Klausen & K. Ståhlberg (Eds.), *New public management i Norden* (pp. 32–49). Odense Universitetsforlag.
Knudsen, T., & Rothstein, B. (1994). State building in Scandinavia. *Comparative Politics*, 203–220.
Kramer, R. M. (2003). Trust and distrust in organizations: Emerging perspectives, enduring questions. Annual Review of Psychology, 50, 569–598. https://doi.org/10.1146/annurev.psych.50.1.569
Krogh, A. H., & Triantafillou, P. (2024). Developing new public governance as a public management reform model. Public Management Review, 26(10), 3040–3056. https://doi.org/10.1080/14719037.2024.2313539
LO, & OAO. (2012). *Undersøgelse blandt offentlige ansatte i LO-forbund*. København
Nagell, H. W. (2021). *Tillitsreform på Norsk*. Agenda.
Nosrati, E., Strand, A. H., Lien, L., & Seifert, L. C. (2022). *Tillitsreformen i offentlig sektor: Rommet mellom offentlig ansattes skjønnsutøvelse og brukernes behov for likeverdige tjenester*. FOFA.
Nyhan, R. C. (2000). Changing the paradigm: Trust and its role in public sector organizations. *The American Review of Public Administration, 30*(1), 87–109.
OECD. (2017). *Trust in government: Evidence, issues, and action*. OECD Publishing.
Pollitt, C., & Hupe, P. (2011). Talking about governance: The role of magic concepts. *Public Management Review, 13*(5), 641–658.
Regeringen. (2011). Et danmark, der står sammen. https://www.stm.dk/statsministeriet/publikationer/et-danmark-der-staar-sammen/
Regeringen, A., Danske Regioner, FTF, Kommunernes Landsforening, & OAO. (2013). *Principper for samarbejde mellem parter på det offentlige arbejdsmarked om modernisering*.
Regeringen. (2016). *Tillit i styrningen Dir. 2016:51*.
Regeringen. (2018). *Sammenhængsreformen: Mere frihed, tillid og tryghed*. Finansministeriet.
Regeringen. (2021). Hurdalsplattformen. https://www.regjeringen.no/no/dokumenter/hurdalsplattformen/id2877252/

Regeringen. (2022). *Ansvar for Danmark: Det politiske grundlag for regeringen.* Statsministeriet.
Regeringen. (2023). DK2030 Danmark rustet til fremtiden. Finansministeriet.
Regeringen. (2024). *Regeringens ældrereform—Du bliver aldrig for gammel til at have det godt.* Social- bolig- og ældreministeriet.
Six, F., & Sorge, A. (2008). Creating a high-trust organization: An exploration into organizational policies that stimulate interpersonal trust building. *Journal of Management Studies, 45*(5), 857–884.
Sigurgeirsdóttir, S., & Johnsen, G. (2018). Public trust in institutions in pre-and post-crisis Iceland (II): Institutionalised mistrust. In *The return of trust? Institutions and the public after the Icelandic financial crisis* (pp. 151–170). Emerald Publishing Limited.
Siverbo, S., Johansson, T., Bentzen, T. Ø., & Winsvold, M. (2023). On the diffusion and implementation of trust-based management in Scandinavia. *International Journal of Public Sector Management.*
Siverbo, S. (2023). Tillitsbaserad styrning och ledning i kommuner och regioner – Spridning, adoption och implementering (KFI-rapport 180). KFI.
Siverbo, S., Johansson, T., Bentzen, T. Ø., & Windsvold, M. (2024). On the diffusion and implementation of trustbased management in Scandinavia: Cross-country survey evidence. International Journal of Public Sector Management, 37(1), 1–19. https://doi.org/10.1108/IJPSM-01-2023-0020
Sørensen, E., & Bentzen, T. Ø. (2019). Public administrators in interactive democracy: A multi-paradigmatic approach. Local Government Studies, 46(6), 945–968. https://doi.org/10.1080/03003930.2019.1627335
Skirbekk, H., & Grimen, H. (2012). *Tillit i Norge.* Res Publica.
Svendsen, G. L. H., & Svendsen, G. T. (2015). The puzzle of the Scandinavian welfare state and social trust. *Issues in Social Science, 3*(2), 90–99.
Svendsen, G. T., & Svendsen, G. L. H. (2016). *Trust, social capital, and the Scandinavian welfare state.* Edward Elgar.
Torfing, J. (2023). From control- to trust-based governance and management. In *Rethinking public governance* (pp. 135–157). Edward Elgar Publishing.
Torfing, J., Andersen, L. B., Greve, C., & Klausen, K. K. (2020). *Public governance paradigms: Competing and co-existing.* Edward Elgar Publishing.
Torfing, J., & Bentzen, T. Ø. (2020). Does stewardship theory provide a viable alternative to control-fixated performance management? *Administrative Sciences, 10*(4), 86.
Torfing, J., & Bentzen, T. Ø. (2022). The Danish control–trust balance in public governance. In *Public governance in Denmark* (pp. 111–129).
Torfing, J., Sørensen, E., & Røiseland, A. (2019). Transforming the public sector into an arena for co-creation: Barriers, drivers, benefits, and ways forward. *Administration & Society, 51*(5), 795–825.
Tillitsdelegationen. (2019). Med tillit följer bättre resultat – tillitsbaserad styrning och ledning i staten (SOU 2019:43). Statens offentliga utredningar. https://www.regeringen.se/contentassets/792a8511629447d8abf5aa2520161631/med-tillit-foljer-battre-resultat--tillitsbaserad-styrning-och-ledning-i-staten-sou-201943.pdf
Vallentin, S., & Thygesen, N. T. (2017). Trust and control in public sector reform: Complementarity and beyond. *Journal of Trust Research, 7*(2), 150–169.
Van De Walle, S. (2013). The role of trust in public services and public sector reform. In *Handbook de administração pública* (pp. 145–158).
Windvold, M., & Bentzen, T. Ø. (2020). Konfliktparadokset: Samarbeid og konflikt i kommunepolitikken. In A. Røiseland & S. Varbo (Eds.), *Folkevalgt og politisk leder* (pp. 34–61). Cappelen Damm.

Tina Øllgaard Bentzen is an Associate Professor at Roskilde University's Department of Social Sciences and Business and a member of the Roskilde School of Governance. Her research explores how public organizations navigate tensions between control and trust, particularly in the context of New Public Management reforms and emerging trust reforms in Scandinavia. Her work examines how trust is adopted, translated, and diffused as a reform concept across countries and organizations. In addition to trust-based leadership and organizational professionalism, she also studies co-creation—both of policy, public services, and governance solutions—focusing on how collaborative approaches can enhance democratic legitimacy and innovation capacity. Tina aims to bridge academic theory and practice, and she frequently collaborates with ministries, municipalities, and professional bodies. Her recent research also investigates how trust in leadership influences outcomes such as innovation, employee motivation and organizational adaptability during times of crisis, such as the COVID-19 pandemic. She is a regular contributor to public debates on governance and public leadership.

Marte Winsvold is a political scientist whose research focuses on institutional trust, democratic legitimacy, and governance reforms within the Nordic welfare states. She works as a research professor at the Institute for Social Research in Oslo and holds a PhD in political science from the University of Oslo. Her work examines how structural changes—such as municipal mergers, participatory arrangements, and trust-based management—shape citizens' trust in political institutions, as well as how politicians perceive control and legitimacy in changing governance contexts. Drawing on a combination of survey research, document analysis, and qualitative interviews, she explores the interplay between policy reforms, media narratives, and public attitudes. Her research contributes to a deeper understanding of how trust is built, maintained, or challenged in the face of institutional and societal change.

Part IV
Security and Foreign Relations

Chapter 7
Nordic Understandings of Security: Commonalities and the End of Exceptionalism?

Niklas Eklund and Monika Gabriela Bartoszewicz

Abstract This chapter scrutinizes ways in which ideas about societal security and total defence contribute to elements of exceptionalism in Nordic security policies. It also traces the geopolitical and historical roots of a transformative policy shift within the region during the early 2020s which followed the unprecedented alignment of Sweden and Finland with NATO. In so doing, the chapter discusses aspects of both inner diversity and regional exceptionalism within Nordic security choices. Whereas many of the intricate and traditional security relationships have continued between the five Nordic states, this chapter contends that new and shared perceived threats have led to a rapid recalibration of their national defence strategies. Whether in response to strategic change in the Arctic or aggressive warfare in East Central Europe, the Nordics are being driven toward both expanded cooperation among themselves and into closer alignment with NATO. Ultimately, however, small but significant differences in their understanding of security remain among them even amidst such accelerated global changes.

Keywords Geopolitics · Security · Nordic balance · Threat perceptions · NATO

7.1 Introduction

What sets Nordic understandings of security apart from those of other countries and regions of the world? Some might argue, not much. Governments in the Nordic countries calculate their economic and geopolitical gains and losses, by and large, in the same way as other governments do, by playing on the strengths and weaknesses of their respective societies. Pragmatism and national security concerns lurk behind

N. Eklund (✉)
Umeå University, Umeå, Sweden
e-mail: niklas.eklund@umu.se

M. G. Bartoszewicz
The Arctic University of Norway, Tromsø, Norway
e-mail: monika.g.bartoszewicz@uit.no

D. C. Nord (ed.), *Contemporary Nordic Public Policies*, Springer Polar Sciences,
https://doi.org/10.1007/978-3-032-11292-7_7

their diplomatic efforts at positive international branding (Coetzee et al., 2024). For the Nordic countries as a group, security policies have for some time been built around certain central values and the projected images of the good society. As described in other chapters of this book, these ideas about social and economic justice, gender equality, and citizens' rights have underpinned far reaching progressive policy making in many areas within the Nordic countries. They have also influenced how national security policies are created. Indeed, before the 2020s, many social scientists in the region would analyze Nordic defense policies not only in terms of how they related to military alliances and geopolitics, but also regarding how they connected with economic and legal structures such as the European Union (Gebhard, 2020).

It could also be argued that the Nordic countries have been singularly advantaged since the end of World War II by enjoying peace in their own region while developing their welfare states. As such, they have been less touched by the harsh realities of geopolitical change. This has also led the Nordics to downplay the role of major international actors in many functional and regional areas related to Nordic security. This is particularly the case with respect to the Arctic and the European High North. There Russia is also an immediate neighbor and significant power.

Of equal interest is what sets Nordic understandings of security apart from the outlooks of other European countries of similar size and levels of development. If by security, we mean "something" which should be valued above all other things (Hoogensen Gjörv et al., 2014), is it possible to talk about a common Nordic understanding of security? Is it possible to identify what the "something" is all about? In their seminal work on the concept of Nordic "societal security," Larsson and Rhinard (2021), suggest there are signs of both convergence and divergence in this vital dimension of security. Their main finding is that the idea of "societal security" is the "something" that should be valued above all other things in security policy. This is a commonality found among the Nordics. Although there are recurrent national debates in all the five countries as to the exact definition and scope of the concept, it is nonetheless an expression of a common core value regarding security that is shared among all the Nordic countries.

Within these national debates, societal security is sometimes referred to with somewhat different words and expressions. It is variably talked about as "comprehensive security," "total defense," "resilience," or as a "whole-of-society approach to security." The wording depends upon which Nordic country and what era one considers. The important point here is the core concept: that the whole population and the totality of national resources are involved in national security and defense. This provides the baseline for all discussion about national security issues among the Nordics. Security in the Nordic countries is also seen as the concern of all citizens and social groups. It is a matter for everybody to think about and to be involved in, contributing to a viable and sustainable national defense, should the need arise.

From an external point of view, these shared ideas of societal security arising from states with a long history of internal and regional strife, war, and domination, may seem a bit strange, perhaps even untrustworthy. Yet, it is there. The concept has evolved into a cornerstone for Nordic understandings of security and how security policies in the five different countries are formulated and legitimized. The intention behind Nordic security policies is to protect and defend all of society. Therefore, all of society should be involved and contribute to the effort in both peace and war. However, this Nordic consensus remains on a conceptual level of understanding and expression. It does not infringe upon the sovereign right of each of the five Nordic governments to define exactly what the words "all" and "society" mean. Nor does it serve as a guideline for common organizational practices or shared understandings when it comes down to defining what or who is a threat to the nation. Thus, there exists a degree of inner diversity within this distinctive regional understanding of security.

Whether referred to as societal, total, or comprehensive defense, the Nordic understanding of such efforts involves the full spectrum of civil and military actors. Safeguarding the economic and social functions of civil society is as important to national security as military capabilities. Public and private actors in the Nordic countries are expected to operate cooperatively in support of the government and the normal functions of society, while at the same time supporting all defense needs. This is a tall order, since the complete protection of the population of any country under modern conditions of warfare or terrorism is impossible. The idea, however, does bring many different types of social, economic, and political actors within the Nordic countries together under one conceptual umbrella. The typical Nordic vertical organization of government structures is complemented by horizontal networks and partnerships when discussing societal security. This involves a wide array of social and economic actors that forge points of contact, planning, and discussion on issues of security.

Moreover, this Nordic usage of terms like societal security, comprehensive security, or total defense can complicate discussions about what are the main drivers and safeguards of national security. In a way, they confound the normal distinction between "hard security" (Mearsheimer, 2001) and "soft security" (Nye, 2004). The traditional hard security perspective tends to emphasize policy measures such as military force, exerting control over economic and social actors, intervening in other countries' affairs, and generally putting pressure on a perceived enemy. Hard security provides a common-sense interpretation of security, which is to avert military threats and to guarantee a modicum of stability through the manifest will and resources of a society to go to war and fight. Soft security, on the other hand, is about persuasion, diplomacy and cultural influences. It stresses cooperation, goodwill, and economic and environmental resilience. Soft security speaks to a more progressive agenda regarding how modern societies can interact, find common ground, and exert global influence without the tragedy of war. A country can become a leader, for example, in promoting new forms of energy extraction that significantly reduce its climate footprint. In a world guided by soft security policies, a country assumes a leadership role, and an international standing, by becoming a model that other countries may wish to emulate. Given these traditional distinctions between different forms of security, the Nordic emphasis on societal security and total defense enlarges and complicates the usual discussion of security choices. Not to put too fine a point on it, but is total defense a call to arms or is it a template for civil resistance? Most Nordics will probably argue that it is a little bit of both.

There is another pillar of the Nordic understanding of security which comes to the fore here—trust. Trust in these societal networks, and trust between government and the citizenry, are the other significant commonalities found among the Nordics in their consideration of security (Jore, 2021). The element of "trust" in national security is probably what makes the Nordic countries different from a lot of other countries and regions of the world. It fosters a certain kind of Nordic "exceptionalism." Among other things, it gives an emphasis to integrative and socializing measures, for example, when it comes to handling terrorist threats. The significant levels of trust in government which underpin other Nordic progressive areas of policy making are assumed to be transferable also to issues of national security. Again, from an international perspective, this too might be a surprising element of their policy development. For instance, when compared with the homeland security approach of the United States, the Nordic interpretation of security would seem to be less focused upon the "inherently violent and uncontrollable" aspect of terrorism (Larsson & Rhinard, 2021, p. 227) and more directed toward integrative or socializing issues. Whereas the U.S. homeland security approach emphasizes popular fear as the rationale for citizen participation in security measures (observing, suspecting, reporting), the Nordic approach is more reminiscent of the trope from British World War II propaganda; "Keep calm and carry on." Nordic citizens are expected to observe and report just as American citizens, but with the understanding that if they do so, their government will deliver both long- and short-term solutions to the problem at hand. Societal security seeks to integrate long-term security measures with existing economic and social policies and approaches.

What policy makers think about security is sometimes separate from what they do. Up to this point, we have touched upon some elements of exceptionalism among the Nordic countries regarding their conceptual understanding of security. However, their actual policy practices deserve further insight. For example, anybody looking at the Nordic countries during the Covid-19 pandemic might have initially assumed a commonality in policy making among the five different countries. Instead, different, individualized and homegrown varieties of lockdown and response were in evidence in nearly all these countries during this period. Borders were closed by most of the Nordics, but at different points in time. Likewise, Sweden was lambasted in the international media for following its own route to achieve collective immunity to the rampant virus. Here, the Nordics were not that "exceptional" in either the sense of being a model to be emulated or in their pursuit of similar policy approaches. The desire to follow a common path in dealing with the pandemic was inherent in all the Nordic governments. Yet this time, they chose not to coordinate with each other. It makes sense, therefore, to take a closer look at each of their policy practices over time to form a better understanding of what constitutes exceptionalism and inner diversity with regard to the broad features of security among the Nordics. This can be begun by considering the specific contexts from which they emerged.

7.2 Geopolitical Realities and the "Nordic Balance"

There are both geopolitical and historic developments that explain how the Nordic region became emblematic of certain soft security approaches–extolling issues of integration, socialization, and welfare politics within international relations. For decades, the Nordic understanding of societal security was praised throughout wide areas of the world as a worthy of emulation. Yet it was rooted in a very secure corner of Europe where a particular Nordic security balance seemed to operate. Most observers today, looking backwards, suggest that it was not that exceptional in character nor necessarily observed by all of the Nordic states in the same way. Regardless, the geopolitical and historical influences that led to this perceived stability and peace in the Nordic region are important factors to consider when seeking to understand overall Nordic views of security today. So too are the specific regional contexts in which they have evolved.

With regard to the latter, two areas have particular relevance—the Arctic and neighbouring areas of Europe. All the Nordic states have national borders within the Arctic. This is why the Nordics are sometimes referred to as the European North, the High North, or the European Arctic. Spanning vast expanses of sea and land in the Northern hemisphere, the Arctic region borders on Canada and the United States as well as Russia. This geopolitical fact makes it hard for each of the Nordic states to ignore any change in the Arctic, whether this is related to climate change, new opportunities for transportation and industry in the region (Dodds & Nutall, 2016), or military threats posed by a resurgent Russia (Östhagen, 2021). There is also a link between Arctic security concerns and home affairs for each of the Nordic governments. The livelihoods of significant portions of Nordic citizens may be affected by Arctic change (Francis & Vavrus, 2012). Similarly, environmental concerns in the Arctic are always high on Nordic agendas. The Arctic policies of the Nordic states also tend to be somewhat elastic, allowing these countries to present different elements of their involvement in the Arctic to different international audiences (Eklund & van der Watt, 2017).

To the immediate south of the Nordics is Central Europe. Further to the east is the East European plain, across which armies have marched back and forth throughout European history. If Russia's attack on Ukraine in 2022 is anything to go by, national armies seem determined to continue doing so into the future. Eastern Europe connects with the Nordic countries to what is commonly referred to as the Baltic Sea Area. This area includes countries such as Russia, Estonia, Latvia, Lithuania, Poland, and Germany. It offers another zone of possible conflict and cooperation for the Nordics. Nordic understandings of security have been shaped by these two geopolitical realities since time immemorial, forging different and evolving constellations of friends and foes in the European High North and the Eastern European region.

However, to understand the security dimensions inherent in the frequently used expression of a "Nordic balance," we need to look further back in time to the Cold War. Were the events and turns in Nordic security during that era an expression of regional exceptionalism or inner diversity? For some time, Nordic behavior during this period was considered by many observers to be "exceptional." In the aftermath of World War Two, the Nordic countries found themselves straddling a hot security border in the emerging bipolar world order. The post-war peace brought a deeply divided Europe with an Iron Curtain between East and West. From a security perspective, the West was orchestrated by the United States, and the East was directed by the Soviet Union. As this bipolar security order congealed, the Nordic states seemed wedged between the two. Their subsequent ability to successfully manage the political and military pressures coming from both sides, gave rise to the notion of a *Nordic balance* (Schöpflin, 1966) within security and defence planning. This concept reinforced the traditional idea of Nordic exceptionalism in the form of a successful regional balancing in response to great power interests. The term soon became synonymous with the longevity of peace and stability in the Nordic region during the second half of the twentieth century (Noreen, 1983; Steinbock, 2008).

International relations in the region also remained largely peaceful both before and after the fall of the Berlin Wall in 1989 and the break-up of the Soviet Union in 1993. They continued to be so during the subsequent expansions of both NATO and the European Union. During all this time, the Nordic states appeared to hold some sort of special key to successful security policy management in an otherwise rapidly changing and increasingly conflictual world. For decades, the idea of a "Nordic balance" seemed to prevail in the minds of those who observed the region despite the significant geopolitical change taking place around it.

In reality, however, this "Nordic balance" was built upon a significant degree of inner diversity among the countries of the region. Denmark, Iceland, and Norway joined NATO in 1949. Finland and Sweden at the same time opted for two different varieties of neutrality. The one chosen by Finland was declared as neutrality underpinned by a pact of mutual recognition and cooperation with the Soviet Union. The Swedish version was based on self-proclaimed non-alignment in peacetime *aimed* at neutrality in the event of war.

Such inner diversity in terms of security policies among the Nordics arose from their different experiences of World War II. Denmark and Norway suffered most of the war under German occupation. Finland fought its way through two separate conflicts with the Soviet Union. It resisted invasion from its eastern neighbor but suffered considerable human and infrastructural loss. Sweden, while carefully and quietly supportive of their Nordic neighbors' plights, remained neutral throughout the war. Iceland, it should be noted, was occupied by British and American military forces. Each of these separate national experiences led to different security choices with the advent of the Cold War (Schöpflin, 1966) and brought with them some diversity in security approaches and policies.

The post-war national security policies that emerged fostered diverse relationships between the Nordics themselves and with the broader international community. In the 1970s for example, Sweden under the leadership of Prime Minister Olof Palme played an active role in trying to create a non-aligned movement among nations unattached to either NATO or the Warsaw Pact. Given that Denmark, Norway, and Iceland were at the time bound by NATO membership, and Finland by its separate agreement with the Soviet Union, Palme's foreign policy move can truly be seen as being "exceptional" within the Nordic region. However, to regard it somehow as a broader Nordic orientation or a common Nordic exceptionalism in foreign relations vis-à-vis the rest of the world, seems problematic.

As observed by Haugevik and Sending (2020, p. 117), there has been a continuing inner diversity between the foreign policies of the Nordic countries throughout their histories—including the Cold War. This condition has been see by some as a "narcissism of small differences." Yet its continued existence reflects a central reality. Despite many similarities in the foreign policy goals of the Nordics, there has been little actual evidence of true policy convergence between the five countries in the area of security during the Cold War or even now. It should be noted, however, that from the era of the supposed "Nordic balance" onward, Nordic governments have tended to extoll the virtues of their "regional cooperation" to international audiences. They have branded themselves and their countries as offering exceptional international role models in this regard. Yet in practice, they have often made actual policy choices with a certain national "identity-defining quality" (Haugevik and Sending p. 110) that sets them apart from one another. They have pursued their own priorities as much as they have contributed to common approaches. (See Chapter 8 in this volume) As such, they have been subject as much to the influences of inner diversity as regional exceptionalism.

With this as background, it is probably correct to argue that the "Nordic balance" model of security policy coordination among the five *appeared* to operate simply because it was never seriously challenged. The actual disruptive elephants in the political room, however, were Finland's and Sweden's varieties of neutrality. Militarily, the Nordics were practically unable to form a single bloc. From the 1950s onwards, Denmark, Norway, and Iceland had to live with the potential reality of an actual alliance conflict with the Soviet Union. Sweden and Finland did not. Neutrality, whether self-styled as in the case of Sweden, or negotiated as in the case of Finland, made an overall collective defense of the Nordic region a problematic concept. This was not because Sweden and Finland underspent on their respective military preparations, or refrained from intelligence operations, but because close coordination among all Nordic forces was not possible. From the perspective of the Nordic NATO members, Sweden and Finland in the "Nordic balance" era could be trusted only to defend the Baltic Sea area and certain land and air routes to the European Arctic. Despite increasing levels of military cooperation between NATO and Sweden and Finland, actual military coordination among the Nordics could only go so far in terms of planning, resourcing, and pre-positioning. For a long time, NATO membership for Sweden and Finland was considered a distant possibility in some faraway future.

We know today that the "exceptional" Nordic balance remained stable well into the twenty-first century. Had the Nordic balance been seriously challenged during the Cold War era a different interpretation of Nordic security policy and actions might have emerged. However, this did not take place, and the concept was largely accepted without scrutiny. The fall of the Berlin Wall in 1989 and the gradual withering away of the Soviet Union in 1991–1993 removed significant threats to Nordic regional peace and seemed to put to rest past geopolitical realities and threats from East Central Europe and the Arctic. Or so it seemed at the time. The "Nordic balance" had survived the Cold War in Europe. The Arctic was deemed to be a "zone of peace." The new societies of Eastern Europe and the Baltics were aspiring to new liberal freedoms and economic development. The pressure on the Nordic countries to align themselves within formerly hostile security environment gradually fell away. In effect, "hard security" considerations in their own region were largely ignored by the Nordics well into the 2000s. It took a sudden resurgence of Russia along with new global economic shifts to bring them back. It was only in 2008–2009 that the five Nordic governments finally agreed to start taking concrete organizational steps toward improved civil and military interoperability (Westberg, 2017). This significant shift in orientation will be considered in the next section of the chapter.

7.3 NATO: Does the Bell Toll for Nordic Diversity in Security?

It seems clear that the Nordic countries, bound by their geographical and cultural proximities to the rest of Europe and the Arctic, have historically charted unique paths within this shared geopolitical reality (Efjestad, 2019). In modern times, their interaction with NATO and other alliances has been deeply influenced by their distinct historical legacies and their separate political circumstances (Tiilikainen, 2006). Denmark, Norway, and Iceland occupy a unique place in this setting, as founding members of the NATO alliance. Their early commitment underscored their post-World War II recognition of the necessity of a collective defence arrangement across the entire North Atlantic area (Archer, 2005).

Their steadfast engagement with NATO spans decades and extends well beyond mere membership status. Denmark, Norway, and Iceland have been unwavering participants in most NATO activities, contributing to the missions, exercises, and operations both within and beyond the territories of the alliance (Græger, 2018). Both Norway and Denmark have made their territories available to host NATO military installations. They have also supported a tradition of joint training exercises with their alliance partners. This level of involvement reflects Norway's and Denmark's dedication not only to NATO's operational prowess, but also their commitment to NATO as a cornerstone of their own national policies aimed at regional security and collective defence.

Iceland, on the other hand, represents a remarkable outlier among the three original Nordic NATO members. The country, albeit a founding contributor to the alliance, remains even today without military forces of its own. Instead, it allows the forces of other alliance members to enter and use its territory and waters. From the very beginning, Iceland has been completely reliant on NATO's security guarantees. This is a testament to the vital role that collective defence arrangements can play even for a nation with limited or no defence capacities of their own.

By contrast, Sweden and Finland have been neutrals in the Nordic region who have adopted a pragmatic approach to their Nordic neighbors, characterized by a delicate balance between cooperation and non-alignment. Their policies have been deeply rooted in their respective national histories and experiences of conflict, particularly in WWII. They emphasize non-alignment as a means of preserving their sovereignty and as their main contribution to regional stability. It should be reiterated, however, that neither country defined their non-alignment in absolute terms. Comparing Swedish and Finnish neutrality with that of Switzerland, for example, suggests significant elements of self-styling and allowing diplomatic back doors being left open by the two Nordics. During the Cold War, neither Stockholm nor Helsinki could avoid discussions with the two superpowers or their allies. Thus, they treaded a fine line between openness, transparency, and partisanship. In terms of trade, Sweden pursued wholesale economic and social integration with the West. Finland, although burdened by its forced economic agreements with the USSR, never wavered in its ambition to eventually follow suit. This parallel strategy of political-military non-alignment and socio-economic integration with the West eventually bore fruit when both Sweden and Finland were admitted to the European Union in 1995. This signalled a significant policy shift in their approach to their security and confidence building measures.

Beginning also in the 1990s, both Sweden and Finland also actively engaged with NATO through the so-called Partnership for Peace (PfP) program, a conduit for cooperation and dialogue between NATO and non-member countries. For a couple of decades, the PfP fostered a degree of transparency and trust, that allowed both Sweden and Finland to collaborate on specific security matters with NATO members while retaining their full sovereignty along with their respective national defence doctrines. Going beyond PfP, in the 2000s, Sweden and Finland became integral contributors to NATO-led military exercises, joint initiatives, and cooperation mechanisms. Their active involvement showcased a pragmatic approach to security cooperation. Both nations recognized the benefits of partnering with NATO, while upholding their non-aligned status (Alberque & Schreer, 2023).

To further solidify their security ties with NATO, Sweden and Finland signed several additional bilateral agreements with the alliance in the early 2000s. These formal frameworks enhanced cooperation on defence and security matters, signifying a commitment to deepening collaboration and information sharing. At the same time, it allowed them to move closer to their other Nordic neighbors—already established NATO members—allowing for a future convergence with them in regional security outlooks. It was, however, the Russian invasion of Ukraine that precipitated their eventual membership within the organization.

7.4 The Rediscovery of a Common Enemy

In the summer of 2022, Sweden and Finland handed in their applications for NATO membership. The governments of the two nations promised to go through the process "hand in hand". However, despite this joint commitment and actual negotiations aimed at joining the alliance together, Finland's accession process proved to be much quicker. Finland joined NATO on April 4, 2023. Sweden, on the other hand was kept waiting due to unfinished ratification procedures in Turkey and Hungary. As repeatedly stated by President Erdogan, the feeling in Turkey was that the Swedish government was too soft on terrorism. In the eyes of Prime Minister Orban in Hungary, Swedish attitudes toward Hungary's own domestic variety of democracy were not supportive and thus its application merited further deliberation in the Hungarian parliament. Despite all the requisite agreements having been signed, Sweden was kept waiting until March 7, 2024. This was in stark contrast with the seamless way Finland joined NATO. However, with their eventual joint membership in NATO, decades of security divergence among the Nordics came to an end.

How can this sudden shift in defence policies among the Nordic countries in 2022 be understood? Almost overnight, all traces of the old "Nordic balance", based on inner diversity in Nordic defence doctrines, was swept away and replaced by a common commitment to NATO membership. It is hard to overlook the profound influence that external drivers and geopolitical cross-pressures exerted on the Nordics in this context. Notably, the return of a shared common enemy was, perhaps, most influential.

A real threat from the east emerged when Russia launched its full-scale invasion of Ukraine on 24 February 2022. The event sent shock waves all through Europe. For the Nordics, it was a stark reminder of the traditional geopolitical realities of the broader European area and what an authoritarian military power might want to accomplish there. Above all, Russia, under the leadership of President Vladimir Putin, had made it clear to the rest of the world that his country was ready to promote and back up its national interests with military force. For the Nordic governments, despite all attempts by the Russian government to downplay its self-proclaimed "special military operation" in Ukraine, it appeared that Russia had emerged as an enemy.

An event in the Arctic Council in April of 2022 further illustrates this nexus between perceived threats among the Nordics and their subsequent defence realignment. Triggered by the Russian invasion of Ukraine, most members of the Arctic Council collectively decided to boycott a previously scheduled meeting. On 4 April 2022, this pivotal decision reverberated across international politics, particularly since security issues were by statute and tradition excluded from Arctic Council discussions. However, this time and in the light of Russia's attack on Ukraine, it seemed impossible for the governments of liberal democracies, including the Nordics, to ignore their collective security concerns. However, the Russian Federation maintained a view that sharply diverged from those of the other Arctic Council members. The Russian Arctic Ambassador's assertion that "Russia firmly believes that there is no potential conflict in the Arctic" crystallized a profound difference in perspective between Russia and the Nordics. Observing the reality of a full-spectrum modern war right on their southern geopolitical doorstep, the Nordic governments were hard put to look the other way. Significantly, even if they did, there was Russia still ramping up its infrastructure and preparedness in the Arctic. At the time, the potential for spillover from Ukraine to Arctic security was vividly captured in a headline from *Newsweek* which read: "Russia says that the crisis in Ukraine must not reach the Arctic. NATO says it has already done so." (O'Connor, 2002).

It could be argued that the Swedish and Finnish aspirations to join NATO came as rational responses to an evolving security landscape. The Russian Federation's warfare in Ukraine, coupled with its divergent perspective on Arctic security evoked deep-seated concerns on the part of Sweden and Finland about the overall stability of the region. Faced with a new security reality in their part of the world, both nations recognized the imperative of joining a collective defense alliance such as NATO. They recognized the strength inherent in numbers.

This realignment underscored a broader trend, however, that has been previously noted above: the transition from a historically diverse set of Nordic defense policies to an emerging convergence and consensus. The shared perception of a common enemy motivated the Nordic countries to recalibrate their defense strategies and cultivate deeper alignment aspirations within NATO. It could also be said, however, that the move toward a collective Nordic defense started well before the invasion of Ukraine, that Russian aggression simply provided new momentum. Sweden, Finland, and the other Nordics, had long been tied to the West financially, politically, and socially. Now they found themselves in close geographical proximity to an outspoken and militarily aggressive neighbor. The subsequent foreign and security policy turns on the part of Finland and Sweden may have been decidedly sharp, discarding their historically important varieties of neutrality, but hardly surprising given the rapidly changing security landscape. To place them in proper context it may be useful to consider the current security challenges faced by the Nordics.

7.5 A Look at Nordic Security Strategies in the 2020s

Nordic security policies are like those of other countries in that they can evoke policies and government directives in other areas than geopolitical and military security. A complete understanding of what the Nordics may do in NATO now that Finland and Sweden are members would require a parallel reading of policy documents in other areas including foreign trade, infrastructural development and energy policy. Also, as governments change in democracies so do their policy goals. Among the Nordics, the foreign policy of a country is published separately from its defense policy, which, in turn, is presented separately from overarching national security documents. Adding to this complexity is the fact that the Nordic countries also organize themselves differently according to their own political processes and decision-making structures with respect to how core security values are fleshed out in public policy. Searching through all these documents, it is not easy to single out those which speak to absolutes in the security priorities of Nordic governments. Looking for current similarities and differences among them can be puzzling. Again, we are looking for how the "something" called security is framed and prioritized in strategic terms. To what extent are the security policies of the five Nordic countries guided by those geopolitical and historic concepts which we have traced so far?

Looking for actual content, however, does provide some items of note. Perhaps the most striking among these is that a contemporary search for *Security policy* reveals government documents which retain remnants of the Nordic inner diversity that was typical for the "Nordic balance" era. However, a simultaneous search for *Defense policy* yields documents focusing on hard security issues most notably having to do with required policy changes necessitated by Finland's and Sweden's accession to NATO. From the results of this comparative exercise, it could be argued that the long-term security policies of the five Nordic governments are lagging their short-term defense policy statements. The former seems to be products of policy inheritance while the latter have evolved from analysis of hard security issues.

Table 7.1 illustrates how some differences in outlook that remain among the Nordics. Looking at the first two columns, there seems to be a nearly uniform acceptance of the "societal security" model along with trust in the stability of state-society relations that was examined at the start of this chapter. There are only slight differences among the countries. The government publications from Denmark, Norway, Sweden, and Finland give a lot of attention to the complexities involved in implementing and managing societal security and trust issues. Iceland, however, does not (Government of Iceland 2024a, b). Iceland publishes some documents on civil security, but these are not linked to national security. Finland and Sweden each spend upwards of 25 pages fleshing out the details of their understandings of societal security (Comprehensive Security in Finland and Total Defense in Sweden). Iceland's separate national security agreement with the US appears front and center in presentation of security matters. The reason for this difference in approach can probably be attributed to choices made by the Icelandic government in how present their documents. A commitment to societal security and social trust can also be found in Iceland, but not in the security publications of its government.

Looking at the regional orientations presented in the third column of Table 7.1, the several Nordic governments are less unified in this aspect of their security views. Iceland again stands apart from the rest with its very clear emphasis on the Arctic and Atlantic settings. Norway too speaks of the importance of the Atlantic region, but largely in the context of linkages to overall European security (Government of

Table 7.1 Examples of inner diversity in Nordic understandings of security as presented in the strategic outlooks of Nordic governments

Country	Societal security	Social trust	Regional orientation	Friends	Foes
Iceland	No	No	Arctic and Atlantic	US, Nato, Nordics	None
Denmark	Yes	Yes	Eastern European	Nato, EU	Russia, Int:l terrorism
Norway	Yes	Yes	Atlantic and Nordic	Nato, Nordics	Int:l terrorism
Sweden	Yes	Yes	Eastern European	Nato, EU	Russia, China, Int:l terrorism
Finland	Yes	Yes	European	Nato, EU	None

Based on the authors review of documents in November 2024

Norway 2024a). Denmark, Sweden, and Finland stress that Europe is their most significant regional orientation. Denmark and Sweden also make it clear in their publications that they see security in East Central Europe and the Baltic Sea area as major security concerns (Government of Denmark 2023a; Government of Sweden 2024a). The Finnish documents express the view that national security is an all-European affair and a common concern regardless of which part of Europe is under threat. The Danish and Swedish documents are more direct in pointing their fingers at Russian aggression and instability in Eastern and East Central Europe.

The last two columns of Table 7.1 indicate whether the Nordic governments are specific about their friends and enemies in their strategic documents. All the Nordic governments seem to express similar opinions about who their friends are. But they diverge when it comes to who they see an enemy. Iceland and Norway are slightly different from Denmark, Sweden, and Finland, in their reiteration of the importance of the Atlantic and Arctic contexts to their national security. NATO is considered a friend by all the Nordic governments, but only Iceland directly points to the United States as a friend. Iceland also mentions that their fellow Nordics are important partners. Norway also suggests that the Nordic region is an important security context. Denmark, Sweden, and Finland emphasize the European Union as their own family of friends, with Europe as their overarching concern in a geopolitical context.

Based on these security policy documents, the results depicted in the last column of Table 7.1 point to the significant differences in outlook between the Nordic countries regarding who they perceive as enemies. Here Iceland is no longer the odd one out but is joined by Finland in suggesting that no specific country, organization, or alliance is an enemy (Government of Finland 2024a). The Icelandic document is very short, positive, and focused on the legal ramifications of its international security linkages. The Finnish document is focused almost entirely on the functional aspects of its comprehensive security model. On the other hand, the governments of Denmark, Norway, and Sweden point to the threat coming from international terrorism and the need to identify countries supporting such activity. Denmark and Sweden take things a bit further. The Danish government also makes it clear that it sees Russia as a major threat to its national security. The Swedish government echoes this position. Sweden also adds China to the mix for being a driver of increased global competition over important natural resources and trade routes.

Will these examples of Nordic diversity remain, and how significant will they be in an era rapid global change? A similar search of documents, this time looking for Nordic *Defence policy concerns*, brings hard security issues to the fore. Iceland is steadfast, presenting the same documents and similar outlooks as cited above. The other Nordic governments, however, offer completely different policy documents on defense matter with statements and outlooks that show remarkable unity in how this type of security is understood among them. The emphasis on soft security measures found in the earlier set of documents is now replaced by hard security concerns. Almost in unison, the defense documents speak of recent turning points and historic shifts in both global and European security. All of them point to new and accelerating hard security needs, most of which concern the need for more investment in their respective military forces.

Denmark, in a defense policy document published in (Government of Denmark, 2023b) points to how "Russia's invasion of Ukraine has brought war back to Europe [and] has underlined the importance of Denmark standing together with our allies and partners in NATO and the EU." The Swedish government presents a speech to parliament by the Minister of Foreign Affairs in March 2024, in which it is said that "Russia has chosen a confrontational approach towards the EU and NATO. This confrontation needs to be addressed." (Government of Sweden, 2024b) Also in 2024, Norway published a long-term defense plan until 2036, mapping their military investment plans "to increase the volume and flexibility of the Norwegian military to enhance our military posture and the defense of the [NATO] Alliance." (Government of Norway, 2024b) Finland, also in 2024, identifies its foreign and security policies as being for a new era noting "the threat of Russian military operations expanding beyond Ukraine cannot be ruled out. No sudden turn for the better is to be expected even if active military operations in Ukraine were to end." (Government of Finland, 2024b).

Gone are the references to soft security approaches and similar measures among the Nordics. The rediscovery of a common enemy drives the five Nordic governments in the direction of a hard security approach and increased military spending. The security-in-diversity approach that for so long dominated Nordic understandings of security seems to have been wiped away by current NATO and military defense needs and these emerge as a common concerns for all.

It would take a longer and a more fine-grained analysis of these and other government statements to get at the real "something" which Hoogensen Gjörv earlier pointed to as being what is valued above all else in Nordic security concerns. However, the obvious discrepancy between the strategic and more current policy documents suggests something very important. The understanding of security among the Nordic countries has taken a new turn. The contributing factors that underpinned the "Nordic balance" for so long are gone. As all observers of public policy know, change comes slowly in strategic documents but quickly in day-to-day governmental decision making. Whether the significant shift from soft to hard drivers of security thinking among the Nordics will completely replace the old order remains to be seen. There are still hurdles to be overcome in the form securing popular support and commitment to a restructuring of the balance between military and civil protection in some of the Nordic countries. There is, however, no mistake that NATO and the accession of Finland and Sweden to the alliance, are driving the Nordics closer together in their understandings of security.

7.6 Conclusion: Still an Exceptional Balancing Act?

In exploring the historic and contemporary policy choices concerning security among the Nordic countries, an intricate relationship among them emerges. Also, difficult policy choices come to the fore. When Sweden and Finland joined NATO in 2023 and 2024, the whole geopolitical situation in the High North of Europe shifted significantly. It moved from the so-called "Nordic balance" model of the late 1900s, based on inner diversity in regional security policy choices, to a more recent path of security policy convergence within the confines of the NATO alliance. The fact that the Nordics are situated along two strategic borderlands complicates security issues. The forces of Geopolitics are evolving and shifting rapidly, both in the Arctic and in Central and Eastern Europe. And the Nordics are destined to play significant roles in the development of security in both areas. The inclusion of Finland and Sweden in the NATO alliance sends a strong signal to Russia and other global actors of their determination to do so. This chapter shows among other things that all the Nordic countries are facing a significant shift both in how they see themselves and how they must choose their friends and foes in the world. Above all, it traces a shift in how security is understood among the Nordic countries.

If the Nordic countries are to successfully navigate the complexities of security in the twenty-first century, they will have to follow a path that gives due attention to both the forces of change and some of the persistent elements of both exceptionalism and inner diversity found within their region. Going forward, a litmus test for the Nordic NATO countries will be their quick adaptation to alliance policies related to civil preparedness, resilience, and peacetime preparations for war. Likewise, cyber security issues, are blurring the boundaries between war and peace in large parts of the world including the Nordic region. The idea that societal security in any form can be pursued by a government, whether it is called "total" or "comprehensive security," is seriously challenged today by electronic interconnectedness and related transnational dependencies. While NATO membership addresses some current regional security imbalances, it simultaneously ushers in a new era of accelerated change in policy formulation as well as additional layers of security concerns for Nordic policy makers. Such changes necessitate the creation of security policies that align themselves with the nuanced dimensions and challenges that have been discussed in this chapter. Security among the Nordic countries can best understood today as multifaceted in character and taking place within a still evolving policy landscape.

References

Alberque, W., & Schreer, B. (2023). Finland, Sweden and NATO membership. *Survival, 2022*, 67–72.

Archer, C. (2005). The NATO Nordics and the CFSP. In L. Miles (Ed.), *The European Union and the Nordic countries* (pp. 269–280). Routledge.

Coetzee, W., Larsson, S., & Berndtsson, J. (2024). Branding 'progressive' security: The case of Sweden. *Cooperation and Conflict, 59*(1), 86–106.

Dodds, K., & Nutall, M. (2016). *The scramble for the poles: The geopolitics of the Arctic and Antarctic*. Polity.

Efjestad, S. (2019). The Nordic region. In J. A. Olsen (Ed.), *Security in Northern Europe* (pp. 37–48). Routledge.

Eklund, N., & van der Watt, L.-M. (2017). Refracting (geo)political choices in the Arctic. *The Polar Journal, 7*(1), 86–103.

Francis, J., & Vavrus, S. (2012). Evidence linking Arctic amplification to extreme weather the mid-latitudes. *Geophysical Research Letters, 39*(6), LO6801.

Gebhard, C. (2020). Scandinavian defence and alliance policies: Different together. In P. Nedergard, & A. Wivel (Eds.), *The Routledge handbook of Scandinavian politics*. Routledge.

Government of Denmark. (2023a). Foreign and security policy strategy, downloaded November 11, 2024.

Government of Denmark. (2023b). Will and ability to take responsibility. Danish defence and security 2024–2033, downloaded January 10, 2025.

Government of Finland. (2024a). Security strategy for society. Government Resolution downloaded November 11, 2024.

Government of Finland. (2024b). Government report on Finnish foreign and security policy, downloaded January 10, 2025.

Government of Iceland (2024a). National security, downloaded November 11, 2024.

Government of Iceland (2024b). Joint understanding, downloaded November 11, 2024.

Government of Norway. (2024a). The Norwegian defence pledge. Long-term defence plan 2025–2036, downloaded November 11, 2024.

Government of Norway. (2024b). Act relating to national security (Security Act), downloaded January 10, 2025.

Government of Sweden. (2024a). National security strategy, downloaded November 11, 2024.

Government of Sweden. (2024b). Statement of Government policy following Sweden's Accession to NATO, downloaded January 10, 2025.

Græger, N. (2018). Need to have or nice to have? Nordic cooperation, NATO and the EU in Norwegian foreign, security and defence policy. *Global Affairs, 4*(4–5), 363–376.

Haugevik, K., & Sending, O. J. (2020). The Nordic balance revisited: Differentiation and the foreign policy repertoires of the Nordic states. *Politics and Governance, 8*(4), 110–119.

Hoogensen Gjörv, G., et al. (Eds.). (2014). *Environmental and human security in the Arctic*. Routledge.

Jore, S. H. (2021). Countering radicalization in Norwegian terrorism policy: A welfare state approach to societal security. In S. Larsson, & M. Rhinard (Eds.), *Nordic societal security. Convergence and divergence*. Routledge.

Larsson, S., & Rhinard, M. (Eds.). (2021). *Nordic societal security*. Routledge.

Mearsheimer, J. (2001). *The tragedy of great power politics*. W.W. Norton & Company.

Noreen, E. (1983). The Nordic balance: A security policy concept in theory and practice. *Cooperation and Conflict, 18*, 43–56.

Nye, J. (2004). Soft power: The means to success in world politics. *Public Affairs*.

O'Connor, T. (2002). Russia says Ukraine crisis must not reach Arctic, NATO says it already has, 31 March 2022, Newsweek, available at https://www.newsweek.com/russia-says-ukraine-crisis-must-not-reach-arctic-nato-says-it-already-has-1693940

Östhagen, A. (2021). The Arctic security region: Misconceptions and contradictions. *Polar Geography, 44*(1), 55–74.

Schöpflin, G. (1966). NATO and the Nordic balance. *The World Today, 22*(3), 114–122.

Steinbock, D. (2008). NATO and Northern Europe: From Nordic balance to northern balance. *American Foreign Policy Interests, 30*(4), 196–210.

Tiilikainen, T. (2006). Finland—An EU member with a small state identity. *Journal of European Integration, 28*(1), 73–87.

Westberg, J. (2017). Försvarssamarbetet: en ny väg till fördjupat nordiskt samarbete? In B. Sundelius & C. Wiklund (Eds.), *Norden sett inifrån. Det fjärde spårbytet*. Santérus Förlag.

Niklas Eklund is Professor of Politics and Public Administration at the Department of Political Science, Umeå University, and the current Director of the European CBRNE Center. He also serves on the steering committee for the Arctic Centre at Umeå University and on the board of the local twenty-first Century Scandinavian Policy Studies Foundation. In his research, he specializes among other things in Arctic security, geopolitics and Russian studies. He was appointed Arctic Six Chair in Nordic and Arctic Security for 2024–2026, a position from which he is actively trying to promote transboundary Nordic research initiatives driven by geopolitical change and emergent security needs in the European High North.

Monika Gabriela Bartoszewicz is an Associate Professor in Societal Security and safety at UIT The Arctic University of Norway where she leads the Secure Societies research group. She specializes in non-linear and cross-sectional threats in Europe, especially in the context of political violence and securitized migration. She has conducted research and worked in the UK, the Netherlands, Italy, Poland, Taiwan and the Czech Republic. Recently appointed as the Arctic Six Chair in Terrorism Studies and as the Visiting Research Fellow at the Institute of National Defense and Security Research in Taipei, she focuses now on dystopias and societal collapses.

Chapter 8
Exceptionalism and Inner Diversity Within Nordic Foreign Policy: The Case of the Arctic Council

Douglas C. Nord and Paul Schmidt

Abstract Over the past two decades the Arctic has become an increasingly important feature of international relations. With the advent of climate change, the search for additional sources of energy and rare earth minerals, and the emergence of new tensions between the circumpolar states, the region has become a steadily growing focus of global attention. All five of the Nordic countries are Arctic states. However, their roles within the region have not been broadly studied. This essay provides an opportunity to examine the Nordic countries' views of the region and to consider their present and future contributions to it. Within the context of their common membership in the Arctic Council, an effort is made to investigate how their actions in that international body have reflected both shared and divergent attitudes and objectives. Such an examination provides an opportunity to assess the sources of Nordic foreign policy stances in the Arctic region as well as their contributions to international diplomacy more generally. It seeks to provide an answer the question: To what extent can both exceptionalism and inner diversity be seen in such undertakings?

Keywords Arctic Council · Nordic states · Diplomacy · Foreign policies · Exceptionalism · Policy diversity

8.1 Introduction

Today the Nordic states, individually and collectively, do not project a large profile on the world stage. Like many "small states" within the global community, their influence is to be felt largely on the margins of international politics (Baldersheim & Keating, 2015). Their contributions tend to be made in the economic, cultural and

---

D. C. Nord (✉) · P. Schmidt
Umeå University, Umeå, Sweden
e-mail: douglas.nord@umu.se

P. Schmidt
e-mail: paul.schmidt@umu.se

D. C. Nord (ed.), *Contemporary Nordic Public Policies*, Springer Polar Sciences,
https://doi.org/10.1007/978-3-032-11292-7_8

social spheres. The "soft power" that they exert can often be seen in the realms of international organization, peacemaking, and humanitarian assistance along with specific efforts to promote environmental protection, human rights and gender equality within the global arena. In pursuing such agenda items, they have earned the collective reputation as being the "conscience of the globe" (Ingebritsen, 2002). Yet it would be incorrect to assume that the specific foreign policies pursued by these Nordic countries are indistinguishable from one another. While they share certain common perspectives and goals based upon shared regional histories, values, and location, they also display a certain degree of difference derived from distinctive national interests, priorities and traditions (Marcussen, 2017). As such, the foreign policies of the Nordic states present both elements of commonality and distinctiveness with regard to one another.

In this chapter we explore these degrees of exceptionalism and inner diversity within the context of the diplomatic efforts of the Nordic countries in one specific area of the globe—the Arctic. Our particular focus is placed on the roles they have played within the Arctic Council. While Norway, Sweden, Denmark, Finland and Iceland are all Arctic states by virtue of their geographic location, they have not always viewed the circumpolar region from the same orientation and perspective. Nor have they advanced exactly the same priorities within the regional body that provides an international governance framework for the broader Arctic community. Some of these differences are revealed when one carefully examines their general conduct within the organization and then looks more specifically at the leadership roles they have played within it. While there are some overall commonalities in the concerns and behavior of the Nordic members of the Arctic Council, there are also significant differences between them in terms of agendas and approach. These may be reflective of certain underlying diversities within the Nordic foreign policies that often go unexamined. We seek to explore some of these.

We begin our inquiry by looking at the different ways in which the Nordic states have encountered the Arctic. While all see themselves as northern societies with long-established heritages and interests in the Arctic, this "northern dimension" has been embraced differently by each nation. We examine such commonalities and differences with an eye towards a better understanding of how each of the Nordic states views the contemporary Arctic and sees their specific roles in the region today. An effort is made to broadly classify their interests and objectives with regard their participation in circumpolar diplomacy and some of the specific features of the foreign policy goals they pursue there.

We then turn our attention to the Arctic Council itself. We outline its origins and development and some of the common concerns that dominate its agenda. We then consider the reasons why each of the Nordics have been interested in taking part in such an international body. We also examine their particular contributions to its successful operation over the past thirty years and some instances where they have presented a common stance and when, at other times, they have gone in different directions. The reasons behind these examples of exceptionalism and inner diversity on the part of the Nordics in their Arctic diplomatic efforts are explored.

Finally, we seek to draw some conclusions regarding the nature and character of Nordic foreign policy as illustrated by the case of the Arctic Council. We seek to determine some of the forces and factors that lead to both solidarity and division. Our goal is to emphasize the point that like most other "normal" states, the Nordic countries can and do pursue foreign policy interests and objectives based upon specific national perspectives and interests. We take the position that while shared bonds and traditions among the Nordic states exert considerable influence in their foreign policy formulations, this is only part of the story. The other looks to the ways in which each of these nations seeks to present their own distinctive foreign policy perspective and approach even within the shared setting of the Arctic.

8.2 The Nordics and the Arctic

One common characteristic that the Nordic states share is that they are northern societies. This important feature is reflected in their histories, cultures, and societal development. It is remarked upon both domestically and externally in almost all discussions that seek to capture the essence of their communities (Browning, 2007). Yet it is one thing to be viewed as "northern" and another to be called "Arctic" (Shadian, 2013; Shaw, 2023). Today, while most of the Nordic governments espouse the latter concept, the vast majority of their populations would be hard pressed to come to the same conclusion. Arctic is not the descriptive term that most of them would offer in characterizing their identity. This is reflective of the fact that most Norwegians, Swedes, Danes, Finns and Icelanders live in the southern portions of their countries and rarely venture into Arctic lands.

As to their government's stance, while at a rhetorical level they may speak of Arctic concerns and priorities, not all of them consistently look to the region as a major focus of their domestic and foreign policies. With regard to the latter one can witness clear divisions among the Nordic nations. Some like Finland, and especially Norway, have embraced the region as a defining source and characteristic of their national foreign policies. Whether it be in the context of its dealings with other circumpolar states or as a defining feature of its relations with the rest of Europe this "northern/Arctic" dimension is regularly invoked (Heininen & Nicol, 2007). This reflects an overall northern orientation that the two countries have adopted in recent years in framing their foreign policy goals and objectives.

This is not the case for the three other Nordic states. Sweden and Denmark both historically and today have primarily looked southward to the rest of Europe and the broader international community. They have normally not sought to define their foreign policy interest in terms of the Arctic. Danish interest in the region is largely confined to its uncertain ties with Greenland, and by virtue of this colonial link, its status and prerogatives within the circumpolar community. Sweden has been called the "reluctant Arctic nation" largely because it has rarely given much priority to the region viewing the Baltic area, the European Union and the broader international community as more important foci for its diplomatic efforts (Sörlin, 2014). Iceland as an island nation in the midst of the North Atlantic has tended to pursue more of an inwardly focused foreign policy viewing developments in the contemporary Arctic largely in terms of their impact on its traditional maritime interests and fishing rights. Thus, it is clear that the basic directional orientation of each Nordic nation seems to have a major influence on how it views the contemporary Arctic and its interests within it.

Another important determinative factor is whether the individual Nordic state considers the Arctic as a "homeland" or as a "hinterland." Again, there exists divisions among them. Norway, and to a certain extent Finland, view the Arctic as a homeland. They have significant territories within the region and consider the lands, waters and people there, integral parts of their own larger communities. For Norway, this is particularly the case with respect to the Far North of the country and Svalbard (Ulfstein, 1995). As such, protecting these areas and their links to rest of the country has been a vital and continuing element of Norwegian consideration of the Arctic. Sweden, Iceland and Denmark on the other hand have tended to view the Arctic primarily as a hinterland—rich in natural resources and largely linked to the rest of their nation primarily in economic and commercial terms (Huskey & Larsen, 2015). Sweden has always considered its thinly populated Arctic area largely in regard to its abundant mineral and forestry resources. Some efforts have been made over the years to respond to the needs and aspirations of its residents, but Stockholm continues to treat the region largely as an afterthought to broader national domestic and foreign policy concerns (Keskitalo, 2019). Similar attitudes have prevailed in Copenhagen with regard to its northern territories. Greenland and the Faeroe Islands have been viewed in the context of what fishing and mineral resources they can offer and regularly deemed to be "special cases" requiring heavy subsidies from the central government. Iceland, as previously mentioned, sees the Arctic today largely in terms of the extractive fishing resources available there.

A third factor which seems to differentiate the Nordic views of the Arctic is the presence, or in one case, the absence of indigenous populations within their borders. Iceland is largely a European settler community and, as such, has found it somewhat difficult to integrate its national experience with the other Nordic histories of indigenous-settler conflict and cooperation. The other four Nordic societies all share this tradition though their experiences, actions and current attitudes vary somewhat. Nor is it the case that the same indigenous populations are represented in each of their Arctic regions. In Norway, Sweden and Finland the primary group is the Sami. In Denmark's case it is the Inuit of Greenland. Thus while, all four nations can and do engage in domestic and regional conversations regarding indigenous needs and aspirations in the Arctic their indigenous interlocutors are not always the same (Fondahl et al., 2015).

A fourth factor which can contribute to a Nordic nation's view of the Arctic is whether it possesses an Arctic maritime border. This has become an increasingly important factor in recent years as fishing, transport, safety and especially energy resources have become prime concerns of the area. Two of the states, Norway and Denmark (via Greenland) have such borders and consequently are regularly part of international diplomatic efforts to describe and define their limits (Byers, 2009).

Understandably, many of their interests and priorities in the region are tied to these and, in turn, frame their overall views of the area. The three others—Iceland, Sweden and Finland—while possessing long-standing maritime traditions, have been largely excluded from these discussions. This can give rise to certain tensions between the two groups as was the case in 2008 with the announcement of the Illulissat Declaration (Nord, 2010). It can also result, as we shall see, in alternating maritime and terrestrial focused priorities being proposed by different Nordic chairs of the Arctic Council.

A fifth factor that can contribute to different Nordic views of the Arctic is whether the country shares a border with Russia. As a major geopolitical player in the Arctic and in the broader global community, Russia has always figured prominently in the minds of all of the Nordics (Gebhard, 2017). However, for two countries, Norway and Finland, it has been a particularly important determinant of their views of the region as they are immediate neighbors to Russia. During the Cold War era, this reality led to very different foreign policy stances taken by Oslo and Helsinki toward Moscow. Yet both endeavored to maintain a working relationship with their larger neighbor and sought to engage Russia in a variety of cooperative cross-border endeavors especially during the era of détente and within the context of emerging circumpolar cooperation (Bones, 2013). Sweden tended to do the same. Denmark and Iceland, for their parts, have experienced a greater sense of "distance" from the perceived challenge of Russia and this has colored their views of the Arctic. Today, with the invasion of Ukraine and the growing Russian military presence in the Arctic, this distinction has become somewhat less significant.

A sixth and related factor affecting Nordic views of the Arctic focuses on NATO membership. Here there is solidarity in that all of the Nordic countries are now members of this political-military alliance. However, this was not always the case. For a substantial period of time the community was divided in two camps. On one side were Norway, Denmark and Iceland who joined the organization in 1949 and have continued as members since then offering vital military resources, bases of operation, and often top leadership to the alliance. On the other side were Sweden and Finland who remained outside NATO pursuing policies of non-alignment and self-defense. Such an arrangement hindered the development of a common Nordic defense initiative and created a certain degree of disagreement regarding how to best address the security needs of the Arctic (Gebhard, 2017). As is discussed in the preceding chapter of this volume, these differences among the Nordics have been significantly reduced since Finland and Sweden have joined NATO. However, certain contending national perspectives remain and continue to color their respective views of the Arctic.

Finally, membership in another international body, the European Union, may also influence the perspectives of the different Nordic states regarding the Arctic. Again, there is division among those who have joined the EU and those who have continued apart. Three countries, Denmark, Finland and Sweden are members though joining at separate points in the community's evolution (Grøn & Wivel, 2017). Norway and Iceland have remained outside and seem content with such a relationship (Thorhallsson, 2015). This division can be seen as having a potential influence over national perspectives on the Arctic in two ways. The first has to do with a sense of

Table 8.1 Factors affecting contemporary Nordic views of the Arctic

FACTOR	Iceland	Norway	Denmark	Sweden	Finland
Orientation	Inward	North	South	South	North
Homeland/Hinterland	Hinterland	Homeland	Hinterland	Hinterland	Homeland
Indigenous people	None	Sami	Inuit	Sami	Sami
Arctic maritime border	No	Yes	Yes	No	No
Neighbor to Russia	No	Yes	No	No	Yes
NATO membership	Yes (1949)	Yes (1949)	Yes (1949)	Yes (2024)	Yes (2023)
EU membership	No	No	Yes (1973)	Yes (1995)	Yes (1995)

Nordic solidarity. Should a "common Nordic perspective" on the needs and priorities of the region be maintained or should this be abandoned on the part of Sweden, Finland and Denmark in favor of an EU-oriented set of policies and programs? The second has to do with whether these same three countries should maintain their own distinct national perspectives even if an overall EU approach is adopted. The currents surrounding these debates can be felt in contemporary discussions within the different Nordic national capitals, Brussels, and the Arctic Council. Table 8.1 captures these seven factors for consideration. They should be kept in mind as we examine the Arctic policies of each of the Nordic states and consider the roles that each has played in circumpolar diplomacy. With regard to the latter, our attention will be focused on the efforts within one particular body—the Arctic Council.

8.3 The Arctic Council

The Arctic Council is now approaching its third decade of existence. Throughout this period of time all of the Nordic states have been members of the body and played significant roles in its development and operation. It remains the most important forum for Arctic policy discussions and offers a basic framework for international management of the region (Nord, 2016a). Through its meetings and the efforts of its specialized scientific working groups, the Arctic Council seeks to promote collaboration among the eight Arctic states, the indigenous peoples of the region and other interested parties. Overall, it seeks to encourage collective efforts to promote the environmental health and protection of the Arctic and to assist the residents of the circumpolar region in securing their health, economic betterment and their cultural heritages (Keskitalo, 2004). As the Arctic has become increasingly the focus of global interest arising from climate change and the scramble to access energy and other natural resources, the work of the Council has become ever more important and complex.

The Arctic Council emerged in the mid-1990s after a decade-long effort by a variety of governmental and non-governmental entities interested in providing a shared approach to meeting the various pressing needs of the region. It came at a time when the Cold War seemed to be coming to an end and the potential for collaboration between East and West appeared possible (Young, 1998). The Arctic was to be a showpiece of how peaceful and constructive cooperation could now take place between the United States and Russia. It was an initiative promoted by Canada both to provide a mechanism for facilitating such interaction as well as means for promoting its own Arctic identity and aspirations (English, 2013).

During these formative discussions, the possible roles and contributions of the Nordic states were given only limited attention by the Canadians. This was in spite of the fact that the Finnish government had played such a decisive role in promoting the earlier Rovaniemi Process that set the foundation for such circumpolar cooperation. It was assumed by Ottawa that once Russian and American support for its initiative was secured, the Nordics would follow along. The Council was to be the domain of the "Arctic Eight"—but it was cooperation between the Arctic Two that seemed to matter most to the Canadians along with their having a leading role in the organization's founding.

This was indeed the path followed by the Arctic Council during the first few years of its existence. Much of its attention was focused on building a sense of trust and common purpose between the members and bridging lingering Cold War divisions. However, two important characteristics of the new body were to give the Nordic states room to display both the distinctiveness of their views of the Arctic and their respective policies toward it. The first of these was that the work of Arctic Council was to be undertaken on the basis of "consensus" among the membership (Nord, 2016b). Thus, it could not be simply an instrument of only two or three nations. In order to do its business, the Arctic Council would require the contributions and approval of ***all*** of its members including the Nordics. The second important factor was that the chair of the body would rotate among the Arctic Eight. Each member state would have the opportunity to periodically present its distinctive vision and priorities for the region (Nord, 2019). The Nordic states—all five of them—were to make use of both of these provisions, and through them, exert an influence within the body that may have not foreseen at the time of its creation. In so doing, they also demonstrated elements of both exceptionalism and inner diversity within these Nordic chairmanship programs. Interestingly, as well, it has been during several of these Nordic chairmanships that the Arctic Council has been seen as undertaking some of its most challenging and effective work. Table 8.2 provides a summary of the membership of the Arctic Council and a chronology of its rotating chair.

8.4 Nordic Behavior Within the Arctic Council

The Nordic states would make use of the consensus and rotating chairmanship features of the Arctic Council to organize their activities within the body and help set the agenda and direction of its efforts. Collectively and individually, they were able to exert perhaps more influence within the organization than its founders may have

Table 8.2 Members of the Arctic Council and their chairmanships

Canada	1996–1998, 2013–2015
United States	1998–2000, 2015–2017
Finland	2000–2002, 2017–2019
Iceland	2002–2004, 2019–2021
Russian Federation	2004–2006, 2021–2023
Norway	2006–2009, 2023–2025
Denmark	2009–2011
Sweden	2011–2013

anticipated. Put simply, the Nordic states were experienced and skilled in consensus-based diplomacy and made use of their capabilities in this area to direct a good deal of the initial work of the organization (Nord, 2016b). In this regard they were unified and "exceptional" in their approach within the body and the other larger states, Russia, the United States and Canada, often relied upon them to set the appropriate tone for the Arctic Council's deliberations and to bridge differences between members when these appeared. The Nordics also played similar roles within the several working groups that were constituted to address specific research needs within the Arctic.

Even here, however, some national differences can be discerned especially with regard to the overall interest and level of enthusiasm which each of the Nordic states displayed in these undertakings. Two members of the group—Finland and Norway—seemed particularly keen to throw their energies into these tasks. This was in keeping with the "northern orientation" that was characteristic of their broader foreign policies and diplomatic efforts. More specifically, Finland was determined to see that the Arctic Environmental Protection Strategy (AEPS) that it had promoted under the Rovaniemi Process would continue under the new Arctic Council (Keskitalo, 2004). Norway, for its part, saw itself as a leading Arctic maritime state and was equally determined that access to and protection of Arctic Ocean marine life and resources be guaranteed to the coastal states within the region. Both countries having shared borders with the Russian Federation were also particularly interested in making sure that their common neighbor was fully integrated into the work of the Arctic Council and played active conciliatory roles when any divisions emerged between the stances taken by the Russian Federation and those of the United States. Also, having an overall homeland attitude toward the Arctic was a force driving them forward to encourage action on a variety of fronts.

The other three Nordic states, Iceland, Finland and Sweden played more secondary roles within the organization during its first years of operation. They often worked in solidarity with Finland and Norway to advance their specific initiatives, but they did so with a decidedly lower profile and degree of enthusiasm than the other two. Here is significant to note that these three all shared (1) a non-northern orientation in their foreign policies, (2) a hinterland view of the Arctic and (3) were not immediate neighbors to Russia. Each of these factors perhaps had an influence over their common decision to be less proactive than Finland or Norway within the organization.

It is interesting to note, as well, that while performing important functions within the Arctic Council, none of the Nordics immediately assumed the position of chair of the body. This was somewhat unexpected given their consensus-building credentials and reputations. However, it was clear from the outset that Canada was determined to be the "founding chair" and derive whatever authority and status might come with that position (Nord, 2006). The leadership baton was next passed to the United States largely to ensure that that country would remain an active and contributing participant in the ongoing work of the organization. It was only then, that one of the Nordics—Finland—assumed the rotating chair. This came about primarily because Russia felt unable to accept the role due to its weak domestic situation and secondarily because Finland had been campaigning for the job as the chief sponsor of the AEPS. Iceland came next largely because its president and government saw the potential to boost to its circumpolar reputation and, again, because Moscow felt unable to assume the post. Finally in 2004, Russia took on the role.

What is interesting to note here is that not until 2006 did any of the Scandinavian states—is this case Norway—assumed the chair of the body. It was also then that the organization would experience three successive Scandinavian leaders. This proved to be of considerable benefit to the work of the struggling organization. Distinctively, it was announced at the time that Oslo, Copenhagen and Stockholm would promote a common five-part "umbrella agenda" during their consecutive leadership terms (Nord, 2016b). This included commitments to (1) advance further research and understanding of environmental threats to the Arctic; (2) provide direction and focus for the sustainable development dimension of the Council's work; (3) promote an international agreement addressing the challenges of search and rescue needs in the Arctic; (4) elevate the profile of the organization and its work within the Far North and the broader global community; and (5) reform its internal operations and procedures. The three Scandinavian governments promised that by "working together" success in each of these areas could be achieved. Such a joint initiative was unprecedented within the organization and truly "exceptional" in character. It was also in keeping with the tradition of solidarity among the Nordic states in conducting their multilateral diplomatic efforts. As will be discussed below, however, this common commitment did not preclude the emergence of certain real differences in priorities and foci that developed among the Scandinavians during this period and thereafter which reflected both their own distinct national foreign policies objectives and orientations toward the Arctic.

8.5 Recent Arctic Council Chairmanship Programs of the Nordic States

As noted above, each of the Nordic states have demonstrated important leadership roles throughout the history of the Arctic Council. Table 8.3 below provides a quick summary of their objectives and accomplishments during their respective terms. It

Table 8.3 Recent Arctic Council chairmanship programs of the Nordic states

Chairmanship	Priorities in Chairmanship	Accomplishments
First Round of Organizational Leadership		
Norway (2006-2009)	• Integrated Resource Management • Climate Change • Reform of Operations	• Mapping of Marine Ecosystem and of Oil and Gas activities • Arctic Climate Impact Assessment • Offshore Oil and Gas Guidelines • Implementing Umbrella Agreement
Denmark (2009-2011)	• Peoples of the Arctic • International Polar Year • Climate Change • Biodiversity (and four others)	• Guidelines for admission of new Observers • Proposal for Permanent Secretariat • Maritime Research Agreement
Sweden (2011-2013)	• The Environment and Climate • The People • The Seas • A Stronger Arctic Council	• Admission of new Observers • Revitalization of the SDWG • Oil Spill Response Agreement • Arctic Vision Statement
Second Round of Organizational Leadership		
Finland (2017-2019)	• Environmental Protection • Connectivity in the Arctic • Meteorological and Oceanographic Cooperation • Education	• Management Guidelines for Arctic Marine Resources • Connectivity and Education Collaborative Projects • Arctic Meteorological Summits • Bridging rifts between Members
Iceland (2019-2021)	• The Arctic Marine Environment • Climate and Green Energy Solutions • Arctic People and Communities • A Stronger Arctic Council	• SAO Marine Mechanism • MOU with AEC • Arctic Youth Initiative • Arctic Council Strategic Plan

is clear that certain commonalities of purpose are revealed as well as distinctive undertakings among the Nordics. Many of these were calculated and planned, while others emerged as responses to pressures and needs felt by the body itself. The particular efforts of each of the Nordic chairs during their leadership periods are briefly discussed below.

8.5.1 Norway—Providing Leadership, Protecting Energy and Marine Resources

Norway's first chairmanship of the Arctic Council was defined by its determination to provide strong leadership, implement the Scandinavian five-part "umbrella agreement," and to solidify its role as a leading Arctic nation. These efforts were deeply rooted in Norway's northern-oriented foreign policy, its steadfast commitment to Arctic cooperation, and its tradition of working closely with Nordic partners. Those driving forces were twinned with an equally strong need to deepen its knowledge of its domestic energy sources and to safeguard its vital marine resources.

The Norwegian chairmanship prioritized three key areas, with integrated resource management emerging as the dominant theme. A cornerstone initiative was promoting the mapping of large marine ecosystems in the Arctic so as to identify vulnerable areas and facilitating sustainable resource use. This project not only addressed global concerns about the environmental impacts of expanding oil and gas activities in a warming Arctic (Norway's Arctic Council Chairmanship, 2006) but also aligned itself with one of the priorities of the Scandinavian umbrella agreement. Norway's leadership efforts produced tangible results which were celebrated as major milestones at the 2009 Senior Arctic Officials meeting and have continued to remain as integral to the Council's agenda (Arctic Council, 2009).

However, Norway's approach also revealed an inherent tension in its leadership approach. While its chairmanship emphasized sustainability and climate change as guiding principles for the organization, its particular focus on mapping oil and gas reserves underscored Norway's strategic interest in positioning itself as Europe's largest oil producer. Framing the Arctic as a peaceful and resource-rich region essential for future prosperity, Norway sought to balance its environmental ambitions with its own economic priorities (Jansen & Mydske, 2014).

Norway's tenure was especially impactful given the limited accomplishments produced by previous chairmanships—those of Finland, Iceland, and Russia—which had been hampered by domestic concerns (Nord, 2016b). Norway seized the opportunity to rebuild and redirect the Council's efforts by establishing a robust framework for activity. In doing so, the country reaffirmed its commitment to Nordic cooperation and to strengthening the Arctic Council's role as a platform for regional governance. However, Norway's leadership also reflected an alignment of its circumpolar priorities with its own national economic agenda, highlighting the complex interplay between regional collaboration and domestic interests.

8.5.2 Denmark—Seeking Internal Harmony, Responding to External Pressures

The subsequent Danish chairmanship of the Arctic Council reflected a fragmented Arctic focus within its own foreign policy. The Danish chair's agenda featured eight loosely connected priorities (Kingdom of Denmark's Arctic Council Chairmanship, 2009). This lack of a cohesive program highlighted Denmark's dual situational challenge: managing internal political tensions with Greenland and the Faroe Islands while asserting itself within the broader Arctic governance framework.

Central to Denmark's program was an emphasis on the Peoples of the Arctic, prioritizing the needs of local communities, particularly their vulnerabilities to climate change. This focus was partly driven by domestic pressures, notably Greenland's 2008 referendum on expanded self-rule, which intensified local efforts to leverage natural resources toward economic independence. The *Act on Greenland Self-Government* (2009) had granted significant autonomy to the community, creating friction with Denmark over a variety of environmental concerns and new Chinese investments in Greenland's mining sector. Similar aspirations for greater autonomy in the Faroe Islands further complicated Denmark's leadership position in the Arctic Council (Grydehøj, 2020). The Danish chairmanship sought to project a sense of unity under a common "Kingdom of Denmark" leadership banner and by addressing the theme of Arctic peoples' needs which were increasingly expressed in its territories.

Denmark also pursued a resource hinterland approach, reflective of its traditional orientation toward the Arctic, emphasizing sustainable resource development while arguing the Arctic's potential for significant economic growth. Integrated resource management was a shared thread that was continued from the Norwegian chairmanship, but Denmark's achievements in addressing broader environmental issues, such as climate change, were limited in nature (Kingdom of Denmark's Arctic Council Chairmanship, 2009). Instead, Denmark prioritized institutional reform, responding to external pressures from China and the EU to clarify the criteria for Observer status within the organization which both sought. These efforts culminated in the Nuuk Declaration (2011), which established a permanent secretariat in Tromsø, formalized Observer admission guidelines, and introduced the Council's first legally binding agreement on cooperation on aeronautical and maritime search and rescue in the Arctic (Arctic Council, 2011). Each of these initiatives sought to showcase Danish diplomatic capabilities and a commitment to strengthening the Arctic Council as a functional organization.

While Denmark made significant contributions to the institutional growth of the Arctic Council, its other somewhat scattered priorities raised questions about the degree of its northern commitment. Ultimately, the Danish chairmanship sought to balance internal political dynamics with external pressures. It endeavored to advance a joint Scandinavian umbrella agenda but in so doing exposing the real limits of Denmark's fragmented policy towards the Arctic.

8.5.3 Sweden—Playing to One's Strengths, Building a Stronger Council

Sweden's chairmanship of the Arctic Council—the last among the Arctic Eight—reflected a state without solid Arctic credentials seeking to find a role for itself in a region where it lacked a strong northern orientation and deeply-rooted policy commitments (Eklund, 2019). As a state without an Arctic coastline and one which had traditionally looked to the south rather than to the north for its foreign policy objectives, Sweden's approach was to leverage its strengths as a skilled and pragmatic diplomat to gain credibility within the organization. This was evident in its focus on institutional reform during its chairmanship and in its efforts to complete the unfinished parts of the Scandinavian "umbrella agreement" (Nord, 2016b).

Sweden's primary leadership achievements came in developing the Council's organizational framework. It implemented the Nuuk Declaration's mandate to establish a permanent Secretariat in Tromsø, which was a significant step in institutionalizing the Arctic Council. Sweden also shepherded the inclusion of several new Observers within the organization including China. An Observer Manual was also produced. These efforts underscored Sweden's emphasis on creating rules and procedures to enhance the Council's functionality and legitimacy (Arctic Council, 2013) and its diplomtic skills in advancing them. Its various initiatives along these lines culminated in the Kiruna Declaration (2013) which provided the Council's first clear vision statement, reaffirming its role as a major regional actor and endorsing the overall Scandinavian approach embodied by the "umbrella agreement".

Sweden substantial diplomatic skills also facilitated the advancement of an Agreement on Cooperation on Marine Oil Pollution Preparedness and Response in the Arctic (Arctic Council, 2013). However, despite listing action on "The Seas" as a top chairmanship priority, Sweden's own lack of an Arctic marine border constrained its activity this area. This resulted in only minimal tangible outcomes in this area beyond that of advancing the oil spill agreement.

Sweden during its chairmanship also invested heavily in revitalizing the Sustainable Development Working Group (SDWG) of the Council. This effort can be interpreted as reflecting Sweden's perception of the Arctic as a resource hinterland. It sought to promote new sustainable resource development initiatives within the SDWG while balancing these with more traditional calls for environmental protection. This re-orientation was to be continued with additional vigor by Canada and the United States in their subsequent chairmanships.

Unlike some of its predecessors, Sweden adopted an institution-focused approach during its chairmanship rather than one of advancing a specific scientific research-oriented agenda. This was consistent with its historically tentative Arctic engagement. Sweden's chairmanship thus reflected its primary identity as a skilled diplomatic consensus-builder, adept at completing unfinished work and strengthening institutional foundations, while pragmatically employing its limited Arctic credentials.

8.5.4 Finland—Advancing Self-interests, Seeking a Common Cause

The Finnish Chairmanship of the Arctic Council exemplified a nation, with a strong northern orientation and a deep commitment to the Arctic as a homeland, striving to achieve meaningful progress amid challenging geopolitical dynamics. Finland's approach reflected its recognized interests and capabilities, with all of its program priorities neatly aligned with established areas of national strength. Its efforts were further shaped by a desire to rebuild consensus within the Council after the stormy chairmanships of Canada and the U.S., which had introduced significant tensions among the membership.

Finland's chairmanship program emphasized four priorities which highlighted its dual focus on sustainability and environmental protection, mirroring its broader foreign policy goals (Finland's Arctic Council Chairmanship, 2017). Finland also aimed to maintain common cause within the Council by fostering collaboration between the U.S. and Russia during a period when the leaders of both countries, Trump and Putin, expressed increasing criticism of multilateral and environmental protection-oriented organizations. This effort reflected Finland's traditional role as a bridge between East and West, shaped by its shared border with Russia, and a longstanding diplomatic concern for Arctic cooperation. However, these attempts ultimately failed to produce a joint declaration at the Rovaniemi Ministerial, the first time an Arctic Council meeting came to an end without such an agreement (Koivurova, 2019).

Despite this setback, Finland achieved significant progress in other areas. It championed both the Ecosystem Approach to Management Guidelines for Arctic Marine Ecosystems and the Regional Action Plan on Marine Litter, underscoring its leadership role in environmental stewardship (Arctic Council, 2019b). Finland also strengthened the Council's collaboration with external organizations, such as the World Meteorological Organization, which can be attributed, in part, to the Finnish nationality of that organization's secretary-general (Koivurova, 2019).

While Finland's leadership agenda initially raised national expectations that it would primarily advance its own economic interests, its chairmanship efforts, instead, prioritized shared goals within the Arctic community and multilateral cooperation. These undertakings reinforced Finland's commitment to fostering sustainability and climate resilience in the Arctic while upholding the Council's collective mission. They also set an expectation that previously agreed Council priorities would continue from one chairmanship to the next.

8.5.5 Iceland—Joint Cooperation, Catching More Fish

The Icelandic Chairmanship of the Arctic Council centered its efforts on four priorities as seen in Table 8.3. Unlike previous Nordic programs, Iceland's focus on Arctic inhabitants did not explicitly highlight Indigenous groups but instead subsumed their needs within the broader category of "Arctic Communities" (Iceland's Arctic Council Chairmanship, 2019). This reflected Iceland's aforementioned difficulties in integrating its national experience with other circumpolar histories of indigenous-settler conflict. However, Iceland did contribute to advancing Indigenous concerns within the Council. A notable achievement during its chairmanship was the establishment of a Permanent Participants Youth Network (Arctic Council, 2021). This action reflected the Arctic Council's continuing commitment to incorporating Indigenous perspectives despite their limited prominence in Iceland's original set of priorities.

On the other hand, Iceland's national interests were particularly evident in its leadership focus on marine issues. As a maritime nation, Iceland's economy strongly depends on the regional fishery, with the seafood industry accounting for 8.1% of GDP—25% when indirect effects are included (Business Iceland, 2021).This factor influenced Iceland to prioritize initiatives such as the formation of the Senior Arctic Officials' Marine Mechanism for addressing maritime concerns, hosting the first International Symposium on Arctic and Sub-Arctic plastic pollution, and developing web-based Arctic Marine Risk-Assessment Guidelines (Arctic Council, 2021). These initiatives underscored Iceland's commitment to both sustainable marine resource management and its national economic orientation which viewed the Arctic as a resource-rich hinterland.

Despite being only briefly mentioned in its chairmanship program, Iceland also made significant strides in strengthening the operation of Arctic Council. During its term of leadership, the Council signed a memorandum of understanding with the Arctic Economic Council, formalizing their ongoing collaboration on sustainable economic development projects in the Arctic, with special emphasis on blue economy initiatives related to Arctic marine resources (Arctic Council, 2019a).

The Reykjavik Declaration of 2021, despite pandemic-related challenges, also marked a milestone in the Council's long-range planning efforts. It included a long-term strategic plan, extending to 2030. It also defined the shared values, goals, and a vision of Arctic collaboration among the members that would continue to guide the Council (Reykjavik Declaration, 2021). This proved to be invaluable as the uncertainties of the Russian and Norwegian chairmanships resulting from the war in Ukraine soon revealed. Iceland's chairmanship, while centered on its national maritime interests, also advanced Indigenous inclusion and institutional strengthening. This resulted in a balanced legacy that integrated national interests with ongoing circumpolar priorities.

8.5.6 The Current Norwegian Chairmanship

During the Russian chairmanship of the Arctic Council (2021–2023), the institution came to an unprecedented standstill following Russia's invasion of Ukraine. The seven western Arctic states issued a statement declaring that, in response to the "flagrant violation of the principles of sovereignty and territorial integrity," they would pause their participation in the Arctic Council (Government of Canada, 2022). This disruption raised doubts about the Council's viability. Without consensus, the body would not do much of its ongoing work. Even the transition of leadership from Russia to Norway in 2023 remained uncertain until days before the official ceremony (Jonassen, 2023).

The Arctic Council, once celebrated as a model for international cooperation, faced the risk of becoming irrelevant. Observers warned that losing its consensus could undermine the critical research conducted by its scientific working groups.

Likewise, it might potentially pave the way for the development of alternative governance frameworks involving non-Arctic states, such as China, Iran, and BRICS members. This was an outcome that the Arctic states were eager to avoid (Gricius, 2023).

At the outset of its chairmanship, Norway had to once again draw upon its diplomatic skills and its established foreign policy that has been traditionally aimed at maintaining normal neighborly relations with Russia. It made it a priority of finding new approaches to fully integrate Russia into the work of the Arctic Council while continuing its solidarity with its EU and NATO partners. Maintaining Russia's continued interest in the efforts of the Council was highly important for Norway. Unlike its withdrawal from some other international organizations (Andreeva & Rottem, 2024), Russia had maintained its presence in the Council, emphasizing that it still considered it to be an important forum (Edvardsen, 2023), However, Moscow also stated that it might leave the organization if it no longer served its interests (Jonassen, 2024).

In light of these challenges, Norway prepared for its leadership role by presenting a program centered on four priorities: the oceans, climate and environment, sustainable economic development, and people in the North (Norwegian Chairmanship of The Arctic Council, 2023). These items were consistent with themes previously agreed to at the Reykjavik session (the last time the Council had formally met) as well as being championed by other Nordic states. Norway sought to offer a sense of continuity in the Council's efforts despite the challenging international environment. Foreign Affairs Minister Eide emphasized, that the primary focus of its chairmanship would be promoting "stability and constructive cooperation in the Arctic" (Arctic Council, 2023).

Norway adopted a strategy emphasizing institution-building and the importance of keeping diplomatic channels open, reminiscent of Denmark's, Sweden's and Finland's previous work on the Arctic Council. While a NATO member, Norway was not bound by EU regulations that called for isolating Russia. This better positioned Oslo to discuss Council matters with Moscow, especially after Sweden and Finland joined NATO (Holm, 2023). In pursuing this strategy of engagement, Norway argued for continued measured cooperation with Russia within a region it considered to be a shared homeland. Demonstrating the benefits of this northern-oriented foreign policy, Norway successfully negotiated fisheries quotas with Russia in the Barents Sea in both 2023 and 2024 (Edvardsen, 2024).

Inside the Arctic Council, itself, Norway used its "powers of the chair" to engage in informal and written exchanges with Russia aimed at maintaining its activities and emphasized the need for continuing collaboration on pressing climate challenges (Andreeva & Rottem, 2024). In February of 2024, Russia declared it would suspend financial contributions to the Arctic Council unless its full activities were resumed (Spence, 2024). Shortly thereafter, Norway announced the return of working group activities through virtual exchanges, thus enabling real-time communication among researchers despite the continued suspension of ministerial and other official political meetings. Intense consultations by Norway with other member states and permanent participants led to this compromise, ensuring the Council's scientific and environmental agenda would proceed (Arctic Council, 2024a). This pragmatic stance allowed Norway to navigate its dual priorities of safeguarding Western interests while preserving the Council as an effective institution.

However, another potential obstacle remained due to the failure of the Council to meet during the Russian Chairmanship and to formally approve an agenda for the upcoming Norwegian Chairmanship. Faced with this unusual situation, Oslo sought to build its chairmanship plans around the last mutually agreed-upon document of the organization, the strategic plan decided at Reykjavik at the end of the Icelandic chair's tenure.

The absence of formal interaction between Russia and the other Council members also allowed Norway to exercise its "powers of the chair" to facilitate additional cooperation at the working group level that was outlined in its chairmanship program. In May 2024, the Council, through the Emergence Prevention, Preparedness and Response (EPPR) working group conducted a virtual exercise simulating a response to a large oil spill, with participation from all member states. The exercise was widely regarded as a successful demonstration of how cooperation could occur despite the challenging international context. The Norwegian chairmanship also initiated both a revision of the management in Arctic large marine ecosystems and a new forest fire initiative (Arctic Council, 2024b). Such efforts suggest that Norway may have also partially returned to its former targeted leadership style by which it endeavored to advance its specific interests and protect its own natural resources within the context of broader circumpolar initiatives.

The Norwegian chairmanship can be seen to have helped stabilize the Arctic Council in a time of crisis. It skillfully used its diplomatic abilities to keep Russia within the body. It promoted continuity of effort by focusing its chairmanship on themes previously endorsed by the Council. It made use of both informal channels of communication and the "powers of the chair" to keep the organization focused on its work. It also managed within a limited range of options to advance some its own national priorities for the Arctic. To solidify these achievements, Norway has already begun working with Denmark, which will assume the chairmanship in 2025, to ensure that a smoother transition in leadership for the Council will take place (Hoglund, 2024).

8.6 Nordic Arctic Council Chairmanships: A Case of Exceptionalism and Diversity

Examining the Nordic chairmanships of the Arctic Council reveals a dynamic interplay between the dual pulls of exceptionalism and inner diversity. Exceptionalism is evident in the remarkable degree of Nordic cooperation and shared commitment to the institution's continuity that was displayed in each of their leadership terms. The regular handover of tasks and priorities between their chairmanships reflects both the tradition of collaboration among Nordic states and their distinctive skills and experience in leading international bodies. The Scandinavian "umbrella program" exemplifies this type of exceptionalism. The focused efforts of the Nordics to further the Arctic Council's institutional development and viability during times of uncertainty and challenge are also evident. The organization has clearly benefited from the specific "Nordic touch" that five of its members have offered.

At the same time, the actions of the Nordic states during their chairmanships of the Arctic Council have not always been in synch with one another nor entirely altruistic in nature. As sovereign states they have operated from their own visions and understandings of the Arctic and often pursued their specific national interests within the region along with shared circumpolar concerns. This is particularly evident in the cases of Norway and Iceland where their respective national foci on energy resources and fisheries have sometimes guided their main chairmanship initiatives. Similarly, states like Sweden and Denmark, who often see the Arctic as a resource hinterland, became advocates of sustainable development during their chairmanships and deliberately moved this issue to the forefront of the Council's efforts. Likewise, when tensions have appeared within the organization between Russia and other members, it has been Norway or Finland (neighbors of Russia) who have taken the lead in addressing these. Finally, it is clear that much of the institutional-building activities on the part of Denmark and Sweden during their chairmanships were not entirely motivated by a concern over the long-term viability of the Arctic Council. A good deal of it stemmed from the fact that neither state had an established and coherent vision of what they wanted to do in the region when they assumed the leadership role. Consequently, they saw the sorting out internal problems of the organization as a useful alternative to advancing a defined Arctic strategy.

This is not to say that national preferences have always trumped circumpolar priorities in constructing Nordic chairmanship programs. We have noted that while some states like Finland have been quite adept in selecting themes that closely align with national strengths and capabilities, others like Iceland—and to a certain extent Denmark and Norway—have had to abandon national preferences in favor of pan-Arctic needs. Despite not having an Indigenous population of its own, Iceland felt the need to accommodate growing community interest in the goals and aspirations of these groups within its chairmanship program. Denmark had to expand its interests beyond Greenland and the Faeroe Islands when it became evident that the Secretariat and Observer questions could not be held in abeyance. And as we have seen most recently, Norway had to shape its current chairmanship around themes it inherited from the Icelandic chairmanship in its paramount effort to keep the organization functioning despite internal divisions arising from the war in Ukraine. Despite this reality, we have also seen that Norway was ultimately able to sandwich some of its particular national concerns related to the Arctic within this inherited framework.

Thus, when we look closely at the behavior of the Nordic states during their leadership terms in the Arctic Council, we can see evidence of both exceptionalism and diversity. This is particularly interesting within an organization that is itself searching for some degree of greater continuity and common purpose in its endeavors. If the Arctic Council can continue to function effectively after the Ukraine crisis, this result will be brought about by all members realizing the merits of having an agreed upon plan that allows for a smooth transition between chairmanships and prioritizes the long-term needs of the circumpolar community over short-term national concerns. The experience of the Nordic chairs suggests that this can be done. While they should not be viewed entirely as "paradigms of virtue" in this regard, there is much in the Nordic willingness to work together and to emphasize the collective good that can be praised.

References

Agreement on Cooperation on Aeronautical and Maritime Search and Rescue in the Arctic. (2011). https://oaarchive.arctic-council.org/items/9c343a3f-cc4b-4e75-bfd3-4b318137f8a2

Agreement on Cooperation on Marine Oil Pollution Preparedness and Response in the Arctic. (2013).

Andreeva, S., & Rottem, S. V. (2024). How and why the Arctic Council survived until now—An analysis of the transition in chairship between Russia and Norway. *The Polar Journal, 14*(1), 229–246.

Arctic Council. (2009). *Report of senior Arctic officials to Arctic council ministers. Tromsø, Norway. April 2009.* Retrieved from https://oaarchive.arctic-council.org/items/d648175f-f2dc-4f1d-b0ce-3e64f8fb3425

Arctic Council. (2013). *Senior Arctic officials' report to ministers. KIRUNA, SWEDEN 15 MAY 2013.* Retrieved from https://oaarchive.arctic-council.org/server/api/core/bitstreams/ef77b639-b12c-4656-b987-f025007350a2/content

Arctic Council. (2019a). *The Arctic council signs memorandum of understanding with Arctic economic counci.* Retrieved 28. November 2024 from https://arctic-council.org/news/the-arctic-council-signs-memorandum-of-understanding-with-arctic-economic-council/

Arctic Council. (2019b). *Senior Arctic officials' report to ministers 2019.* Retrieved from https://oaarchive.arctic-council.org/items/ef17110b-e043-4a19-a52e-21f066377cb9

Arctic Council. (2021). *Icelandic chairmanship 2019–2021 factsheet.* Retrieved from https://oaarchive.arctic-council.org/items/3703180d-90e4-4476-b2c6-4ff533de5650

Arctic Council. (2023). *Norway's chairship, 2023–2025.* Retrieved December 4, 2024 from https://arctic-council.org/about/norway-chair-2/

Arctic Council. (2024a). *Arctic council advances resumption of project-level work.* Retrieved December 4, 2024 from https://arctic-council.org/news/arctic-council-advances-resumption-of-project-level-work/

Arctic Council. (2024b). *One year into the 2023–2025 norwegian chairship: A Q&A with SAO chair Morten Høglund.* Retrieved December 4, 2024 from https://arctic-council.org/news/one-year-norwegian-chairship/

Baldersheim, H., & Keating, M. (2015). *Small states in the modern world: Vulnerabilities and opportunities.* Edward Elgar Publishing.

Bones, S. (2013). Science in-between: Norway, the European Arctic and the soviet union. In S. Sörlin (Ed.), *Science, geopolitics and culture in the polar region. Norden beyond borders.* Routledge.

Browning, C. S. (2007). Branding Nordicity: Models, identity and the decline of exceptionalism. *Cooperation and Conflict, 42*(1), 27–51.

Business Iceland. (2021). *Icelandic fisheries.* Retrieved November 28, 2024 from https://www.responsiblefisheries.is/media/1/icelandic-fisheries-press-kit-mai-2021-enska.pdf

Byers, M. (2009). *Who owns the Arctic?: Understanding sovereignty disputes in the North.* Douglas & McIntyre.

Edvardsen, A. (2023). *Russia's top Arctic diplomat: We still consider the Arctic council an important forum.* Retrieved December 4, 2024 from https://www.highnorthnews.com/en/russias-top-arctic-diplomat-we-still-consider-arctic-council-important-forum

Edvardsen, A. (2024). *Norway and Russia reached a fisheries agreement for 2024.* Retrieved December 4, 2024 from https://www.highnorthnews.com/en/norway-and-russia-reached-fisheries-agreement-2024

Eklund, N. (2019). The Swedish chairmanship: Foresight and hindsight in Arctic activism. In D. C. Nord (Ed.), *Leadership for the North: The influence and impact of Arctic council chairs* (pp. 71–88). Springer International Publishing.

English, J. (2013). *Ice and water: Politics, peoples and the Arctic Council.* Allen Lane.

Finland's Arctic Council Chairmanship. (2017). *Finland's chairmanship program for the Arctic council 2017–2019.* Retrieved from https://oaarchive.arctic-council.org/server/api/core/bitstreams/cd14f8d1-cba1-4802-ab3e-67ac12b9e544/content

Fondahl, G., Filippova, V., & Mack, L. (2015). Indigenous peoples in the new Arctic. In B. Evengård, J. N. Larsen, & Ø. Paasche (Eds.), *The new Arctic* (pp. 7–22). Springer International Publishing.

Gebhard, C. (2017). Scandinavian defence and alliance policies: Different together. In P. Nedegaard, A. Wivel (Eds.), *The Routledge handbook of scandinavian politics.* Routledge.

Government of Canada. (2022). *Joint statement on limited resumption of Arctic Council cooperation.* Retrieved December 11, 2024 from https://www.canada.ca/en/global-affairs/news/2022/06/joint-statement-on-limited-resumption-of-arctic-council-cooperation.html

Gricius, Gabriella. (2023). *From Russia to Norway: Three scenarios for the Arctic Council's future after the chairmanship handover*. Retrieved December 4, 2024 from https://europeanleadershipnetwork.org/commentary/from-russia-to-norway-three-scenarios-for-the-arctic-councils-future-after-the-chairmanship-handover/

Grøn, C., & Wivel, A. (2017). Scandinavia and the European Union: Pragmatic functionalism reconsidered. In *The Routledge handbook of scandinavian politics.* Routledge.

Grydehøj, A. (2020). Unravelling economic dependence and independence in relation to island sovereignty: The case of Kalaallit Nunaat (Greenland). *Island Studies Journal, 15*(1), 89–112.

Heininen, L., & Nicol, H. N. (2007). The importance of northern dimension foreign policies in the geopolitics of the circumpolar north. *Geopolitics, 12*(1), 133–165.

Hoglund, M. (2024). *Norwegian chairship of the Arctic Council* [Interview]. Arctic Circle. https://www.youtube.com/watch?v=_qy0lTJB0JE

Holm, A. (2023). *Arne O. Holm says the Arctic Council is dead. Long live the Arctic Council.* Retrieved December 4, 2024 from https://www.highnorthnews.com/en/arctic-council-dead-long-live-arctic-council

Huskey, L., & Larsen, J. (2015). The Arctic economy in a Global Context. In

Iceland's Arctic Council Chairmanship. (2019). *Together towards a sustainable Arctic.* Retrieved from https://oaarchive.arctic-council.org/server/api/core/bitstreams/914307b5-3323-45ac-9319-5ac0f99790ca/content

Ingebritsen, C. (2002). Norm entrepreneurs: Scandinavia's role in world politics. *Cooperation and Conflict, 37*(1), 11–23.

Jansen, A.-I., & Mydske, P. (2014). Norway: Balancing environmental quality and interest in oil. In (pp. 181–207).

Jonassen, T. (2023). *Arctic circle assembly 2023: Arctic council after Russia's handover: "We Are Still Here".* Retrieved December 4, 2024 from https://www.highnorthnews.com/en/arctic-council-after-russias-handover-we-are-still-here

Jonassen, T. (2024). *Russia threatens to withdraw from the Arctic Council.* Retrieved December 4, 2024 from https://www.highnorthnews.com/en/russia-threatens-withdraw-arctic-council

Keskitalo, E., & Carina, H. (2004). *Negotiating the Arctic: The construction of an international region.* Routledge.

Keskitalo, E., & Carina, H. (2019). *The politics of Arctic resources. Change and continuity in the "Old North" of Northern Europe.* Abingdon.

Kingdom of Denmark's Arctic Council Chairmanship. (2009). *The Kingdom of Denmark. Chairmanship of the Arctic Council 2009–2011*. Retrieved from https://oaarchive.arctic-council.org/server/api/core/bitstreams/7cd7f8b3-c7a7-467b-b25f-b5fc9f9105b3/content

Kiruna Declaration. (2013). https://oaarchive.arctic-council.org/items/bc9e3845-fb69-40e8-a1da-0b516c4ffbd6

Koivurova, T. (2019). *Did Finland fail as Arctic Council chair?* Retrieved November 28, 2024 from https://www.arctictoday.com/did-finland-fail-as-arctic-council-chair/

Marcussen, M. (2017). Scandinavian models of diplomacy. In Null (Ed.), (1 ed.) Routledge.

Nord, D. C. (2006). Canada as a northern nation: Finding a role for the Arctic Council. In N. Michaud, P. James, & M. J. O'Reilly (Eds.), *Handbook of Canadian foreign policy*. Lexingtion Books.

Nord, D. C. (2010). The shape of the table, the shape of the Arctic. *International Journal, 65*(4), 825–836.

Nord, D. C. (2016a). *The Arctic Council: Governance within the far north*. Routledge.

Nord, D. C. (2016b). *The changing Arctic: Creating a framework for consensus building and governance within the Arctic Council*. Palgrave/Macmillan.

Nord, D. C. (2019). Chairs and International Organizations: The case of the Arctic Council. In D. C. Nord (Ed.), *Leadership for the North: The influence and impact of Arctic Council Chairs* (pp. 53–68). Springer International Publishing.

Norwegian Chairmanship of The Arctic Council. (2023). *Norway's chairship Arctic Council 2023–2025*. Retrieved from https://www.regjeringen.no/contentassets/034e4c4d49a44684b5fb59568103702e/230322_ud_ac_programbrosjyre_en_web.pdf

Nuuk Declaration. (2011). https://oaarchive.arctic-council.org/items/c47f7683-fbc2-41d9-ad79-17577c887e20

Reykjavik Declaration. (2021). https://oaarchive.arctic-council.org/items/4e4d237c-9570-49b3-b815-c22b605e8296

Shadian, J. (2013). The Arctic gaze: Redefining the boundaries of the Nordic region. In S. Sörlin (Ed.), *Science, geopolitics and culture in the polar region: Norden beyond borders* (pp. 259–289). Ashgate Publishing.

Shaw. (2023) ng China threat: The US' securitization of the 'Belt and Road Initiative'. *Chinese Political Science Review, 8*(1), 84–104.

Sörlin, S. (2014). The reluctant Arctic citizen: Sweden and the North. In D. Klaus, & C. Powell Richard (Eds.), *Polar geopolitics: Knowledges, resources and legal regimes* (pp. 149–165). Edward Elgar Publishing.

Spence, J. (2024). *Russia suspends funding for the Arctic council: Wake up call not death knell*. Retrieved December 4, 2024 from https://www.highnorthnews.com/en/russia-suspends-funding-arctic-council-wake-call-not-death-knell

Thorhallsson, B. (2015). The outsiders: Norway and Iceland. In P. Nedegaard, & A. Wivel (Ed.), *The nordic countires and the European Union: Still the other European community?* (1st ed.). Routledge.

Ulfstein, G. (1995). *The svalbard treaty: From terra nullis to norwegian sovereignty*. Scandinavian University Press.

Young, O. R. (1998). *Creating regimes: Arctic accords and International Governance*. Cornell University Press.

Paul Schmidt is a doctoral student in the Department of Political Science and the Arctic Graduate School at Umeå University, Sweden. His research focuses on evolving security dynamics in the Arctic, with a particular emphasis on the emergence of novel threats and responses thereto. His PhD project investigates how the seven western Arctic states address cybersecurity risks, misinformation, and critical infrastructure vulnerabilities, shedding light on how global security challenges are increasingly manifesting in a region traditionally perceived as remote and pristine. He emphasizes that the Arctic is not a homogeneous space, but rather a politically diverse region where distinct national interests are deeply interwoven with broader international developments, giving rise to unique governance and security challenges. Schmidt's work aims to refine and expand theoretical frameworks–particularly securitization theory–to better account for the complexity of contemporary security challenges. He is especially interested in how Arctic governance intersects with hybrid threats and strategic competition. His broader research interests include Arctic security, Arctic policy and hybrid threats.

Part V
Climate Change and the Environment

Chapter 9
The Impact of Climate Change on the Nordics from a One Health Perspective

Birgitta Evengård and Anastasia Emelyanova

Abstract The Nordic countries are like the rest of the world largely affected by the processes brought about by human impacts on the Earth. However, the speed of climate change in the Arctic is several times as fast as in the rest of the world, a phenomenon known as Arctic amplification (Rantanen et al. in Commun Earth Environ 3:1–10, 2022). Thus, climate change in the Arctic is not just having an impact on the Nordic region, but it also acts as a driver for conditions in the rest of the globe. This makes the Nordic North of particular interest to the whole global society. Among the most important changes detected here are declines in nature, directly driven by land and sea use (change, intensification); direct exploitation of resources (especially fishing and mining); pollution; the expansion of invasive species; the effects of globalization and its various economic components; geopolitical tensions; and other significant processes worthy of consideration. This chapter summarizes the rapid development of the One Health approach in addressing the specific challenges posed by climate change and zoonotic diseases in the Nordic region. It highlights collaborative efforts among Nordic countries, including Arctic Six cooperation, and other major regional funding efforts at research and education. It discusses how the utilization of One Health principles can address the needs of the region and noted the importance of interdisciplinary collaboration in achieving identified goals. Also, it discusses the specific health challenges posed by environmental contaminants, the transmission of zoonotic diseases, and the impact of climate change on species distribution that are seen in the Nordic North. It also considers the role of citizen science in data collection and the need for international collaboration in addressing these complex issues. Overall, the chapter underscores the urgency of addressing environmental and health challenges in this area through evidence-based policymaking and community engagement in the Nordic region.

Keywords Nordic countries · Climate change · The Arctic · Health · Well-being

B. Evengård (✉)
Umeå University, Umeå, Sweden
e-mail: birgitta.evengard@umu.se

A. Emelyanova
University of Oulu, Oulu, Finland

D. C. Nord (ed.), *Contemporary Nordic Public Policies*, Springer Polar Sciences,
https://doi.org/10.1007/978-3-032-11292-7_9

9.1 Climate Change and the Nordics

The Nordic countries are significantly affected like the rest of the world by processes brought about by human impacts on the Earth. Among the most important ones are decline in nature, directly driven by land and sea use (change, intensification), direct exploitation (especially fishing), pollution, invasive species, and indirectly driven by globalization and various economic activities, trade, shipping, tourism, unsustainable practices in economic incentives, consumerism lifestyle, geopolitical tensions, and other processes (University of Jyväskylä, 2023). The risks and consequences of such activities often depend heavily on the specific local context. A higher rate of Arctic warming, called Arctic Amplification, is found in the Far North of the Nordic region. When sea ice melts during the Arctic summer, the maritime waters there will absorb more heat from the sun, thus speeding up the melting of the remaining ice. These same kinds of dynamic processes also have an impact on snow-melting, bare-soil absorption of sunlight energy, and the associated thawing of permafrost (Dai, 2021).

The Arctic, including the northern regions of Nordic countries, is witnessing major changes which go beyond historical variations. The risks to health for both ecosystem and society are increasing. For climate and ecosystems, certain tipping points are critical and will determine the speed and possible reversibility of change. Threats to food and health are connected to many of these ongoing and projected changes in climate and ecosystems. Even with ambitious mitigation efforts, many ocean and cryosphere-dependent communities may face challenges to their adaptation limit during the second half of the twenty-first century (van Oort et al., 2022).

Some of the most ancient Arctic permafrost has warmed by more than 0.5 °C since 2007 (Arctic Monitoring and Assessment Programme (AMAP), 2017). The snow-cover season is becoming ever shorter. Between 1982 and 2011, the period of snow-cover in the Eurasian Arctic decreased by 12.6 days/year on average. The North American Arctic lost 6.2 days of snow-cover duration per year during the same period. These changes affect ecosystems and local human populations in a significant fashion (Bokhorst et al., 2016). A transdisciplinary, comparative analysis reveals key risks from Arctic permafrost thaw. These include infrastructure failure, disruption of mobility and supplies, decreased water quality, challenges for food security, and exposure to diseases and contaminants. The study's novelty resides in the comparative approach spanning different disciplines, environmental and societal contexts, and the transdisciplinary synthesis considering various risk perceptions (Gartler et al., 2025).

Most current climate models predict that at the present rate of atmospheric CO_2 increase; the Arctic will be completely free of summer-ice by the 2030s (Arctic Monitoring and Assessment Programme (AMAP), 2017; Jahn et al., 2016). Less Arctic Sea ice means an extended duration of open waters, which, in turn, can attract additional economic activities such as fisheries, energy production, mining exploration, and increased shipping along "the northern Arctic sea route." The warming of the Arctic Ocean, and its freshening and dilution from the thawing of glaciers and sea ice also affects ocean circulation. It decreases the formation of cold, relatively dense, deep-strata water, which, in turn, may weaken the Gulf Stream of the Atlantic Ocean. This may have implications for the global weather system. The frequency of intense hurricanes, heat waves, and wildfires, are now increasing in the northern hemisphere (Samenow, 2021; Schiermeier, 2018).

In the Nordic Arctic, the effects of such change have long been recognized and felt by the people who are living in the region. Climate adaptation has already become part of their ordinary lives (Bokhorst et al., 2016; Hovelsrud et al., 2011). Indigenous peoples such as the Sami in the Sápmi region and the Inuit in Greenland have developed adaptations to their changing environments. Their local, traditional and indigenous knowledge is of inestimable value.

9.2 Nordic Cooperation in International Environmental Diplomacy, Science and Education

Nordic countries were among the first within the global community to call attention to emerging environmental threats. They were behind the efforts to convene the first major environmental conference on the global environment and helped to initiate international negotiations on the issues of climate change, biodiversity loss, environmental degradation, and other pressing threats to planetary well-being. In the late 1960s it was decided on the initiative of Sweden to hold the first United Nations Conference on Environment and Development (Stockholm Conference 1972). In this initiative, Sweden noted how changes in the natural environment had become a pressing problem in both developing and developed countries and that international cooperation was needed to address them. The conference attendees observed that the protection of the environment and development should be seen in relation to each other rather than as separate undertakings. They agreed on the principles directed toward a more sustainable use of global resources (based on materials from University of Jyväskylä, 2023).

The best-known definition of sustainable development is the one produced by the United Nations World Commission on Environment and Development (WCED) in 1987 and chaired by former Norwegian Prime Minister Gro Harlem Brundtland. The Brundtland Commission report noted that "Sustainable development is development that meets the needs of the present without compromising the ability of future generations to meet their own needs." (United Nations, 1987). Nordic countries have continued to play active roles in addressing both climate change issues and sustainable development concerns throughout the globe, but especially in their own Arctic regions as discussed in the preceding chapter of this volume. In 2022, Sweden convened another UN Environment Summit some fifty years after the original one. Some of its major concerns were the topic of planetary boundaries, Agenda 2030, and the new Sustainable Development Goals. The Nordics have also worked among themselves to further regional efforts at research and education related to environmental protection, sustainable development and climate change. Box 9.1 summarizes those participants who have been involved and some of their undertakings.

Box 9.1 Nordic Countries' Cooperation in Research and Education Related to Climate Change and One Health

The effects of climate change (cc) do not respect political or administrative borders. This makes cc collaboration networks between countries imperative. The Arctic Monitoring Assessment Report and the Sustainable Development Working Group under the Arctic Council has successfully dealt with issues concerning the whole Arctic. Nordic science actors have been instrumental operators to support regional cooperation in societally vital topics, including cc and its effects on human, animal, and environmental health (One Health). Some examples of cooperation in the Nordic region with extensions to the Arctic, Baltic and EU areas:

- **Nordic Council of Ministers** (NCM) supports the search for future climate solutions and contributes to the climate agenda via its working groups and institutions such as e.g. **Nordregio, Nordic Institute in Greenland** and funding programmes **Nordforsk, Nordplus** etc. NCM organizes various events, workshops to support the negotiations, the implementation of the Paris Agreement and the UNFCCC, publishing quality reports including Vision 2023
- **NCM** funds projects that must involve at least three Nordic countries plus one or more non-Nordic partners. Since 2012, the **NCM Arctic Programme** funded over 170 development related projects, the 2022–2024 focus being on the four Ps: Planet, People, Prosperity and Partnerships in the Nordic and Arctic regions. The current program period 2025-2027 is being implemented with a focus on creating socially, economically, and environmentally resilient communities
- **NordForsk** established a **NORIA-net** in 2012 to assess the potential for added value in a larger-scale joint Nordic research initiative. The NORIA-net Arctic formed the basis for a joint Nordic research programme, which Norway, Sweden, Iceland, Denmark, Finland and Greenland funded through a common pot. NordForsk established in 2016 four **Nordic Centres of Excellence** in Arctic research
- **Kolarctic CBC** 2014–2020 funding programme was co-funded by the EU, Sweden, Finland, Norway and Russia. Thematic Objectives 6 (Environmental protection, cc adaptation and mitigation) and 7 (Improvement of accessibility to the regions, development of sustainable and climate-proof transport and communication networks and systems) funded nearly 30 projects via 4 calls of proposals. The programming of the 2021–2027 period has been discontinued due to Russia imposing the war with Ukraine

- **Interreg Aurora** 2021–2027 is a key European Union (EU) instrument for cross-border cooperation. The programme offers great opportunities and enables new and exciting cross-border cooperation in the Nordic part of Europe and Sápmi. One of its four priorities is Priority 2—A greener Europe with Specific Objective 2.1 Cc adaptation, on increasing awareness of cc and find methods to be able to prevent/mitigate and adapt
- **University of the Arctic** co-funds projects together with the Danish Ministry of Higher Education and Science, Norwegian funding operated by the UiT – the Arctic University of Norway, and Government of Canada with Memorial University of Newfoundland operating involvement of Canadian partners
- **The Arctic Six** is a partnership between Luleå University of Technology, UiT The Arctic University of Norway, Umeå University, University of Lapland, University of Oulu, and Nord University. This alliance advances knowledge, develops education, undertakes research and creates innovations for the development of the Nordic region and a sustainable Arctic via e.g. Chairs & Fellows program focusing on issues relating also to One health and cc
- A number of other Nordic options are available to support scientific, educational, and cultural cooperation in the region. Read more e.g., here https://napa.gl/en/funds/other-nordic-funds/.

9.3 Health Risks to the Globe, and the Nordics, Arising from Climate Change

As noted earlier, there are growing threats to global ecosystems and to humans as climate hazards grow. Health concerns for affected human populations are particularly pressing. These include matters related to increased heat-related human mortality and morbidity, food-borne, water-borne and vector-borne diseases, and mental health challenges. Animal and human health impacts of climate change in the Arctic are both direct, resulting from changes in temperatures and/or extreme climate events, and indirect, resulting from how climate change affects livelihoods and foods, infrastructure, species ranges, the viability of parasites, bacteria and viruses (de La Rocque et al., 2008; Larsen & Fondahl, 2015). The following discussion focuses on what is being done by the Nordics to respond to such challenges and illustrates how one particular initiative may have an important impact.

9.4 The Arctic Environment and a One Health Approach

The Arctic provides an example of an extreme natural environment, manifested in many climatic, geochemical, and biotic factors. The flora and fauna (including humans) of the Arctic are adapted to a highly seasonal environment with extremes in temperature and humidity. As this landscape changes today, the stressors on the endemic flora and fauna increase and invasive species become more common. The most obvious parameters that can influence the spread of natural infections are temperature, precipitation, and fluctuations in the length of the seasons. Changes in these parameters can affect the suitability of habitats for hosts and vectors, and their reproduction rates, distribution, and abundance. These can have an impact on human health. The existing natural relationships in the Arctic are often complex, and the introduction of socio-economic factors, such as human behavior and land use, can often add to the effects of climate change effects in the region. Such a setting seems ideally suited for a One Health approach, connections and actions in the Arctic (Berner et al., 2024).

The One Health approach to research and education, historically focused on matters related to zoonoses and was initiated and led by the veterinary and disease ecology participants (Castañeda et al., 2023). Today, this orientation has grown to embrace wider connections between human, animal, and environmental health and now draws upon the energies of a wider group of researchers and educators. Understanding these types of connections allows one to respond to health challenges in a holistic way and address problems at their root causes rather than treating their outcomes. One Health addresses multiple sectors of public policy development. In its research efforts multiple disciplines work together, and participants from various levels of society work together to foster enhanced well-being and tackle threats to health and ecosystems (Castañeda et al., 2023; Evengård et al., 2021). A One Health approach is incorporated into all components of the International Health Regulations Monitoring and Evaluation Framework (De La Rocque et al., 2019).

While human and animal health professions have only relatively recently developed and adopted the One Health term, its precepts are not new. One Health, concepts and ideas recognize the interconnectedness of all living beings and their environment. These have been at the core of Indigenous worldviews for millennia (Kutz & Tomaselli, 2019).

9.5 Contaminants and Pollutants as Health Challenges

Although climate change is frequently presented as the most severe trigger for health challenges in the Nordic region, it is neither the first nor the only one within the Arctic area that has had to be confronted. Before climate change, there were other problems, coming mostly from lower latitudes, such as long-range pollution from radioactive contaminants and heavy metals. All were sources of concern to Indigenous and other local peoples, NGOs, and the research community (Heininen et al., 2020). Environmental contaminants, terrestrial and marine toxins and pollutants are a major problem and an essential component of a One Health model.

Even though the northernmost areas of the Nordic region are often far from larger industrialized areas, the environment in the high North still carries the traces of pollution from soot, (micro)plastics, methane, pesticides, marine litter, wood combustion and oil and gas flaring. Many of these contaminants are transported over long distances via rivers, oceans, and the air. They can have far reaching negative impacts on the environment and human health (Arctic Council, 2023).

Since its establishment in 1991, Arctic Monitoring Assessment Program (AMAP) has studied human biomonitoring and exposure, health effects and risks associated with measured contaminants in the Arctic. Nordic researchers have made significant contributions from Nordic researchers including their publishing of data, key findings, research priorities and recommendations in the series of high-quality reports. The first AMAP report was presented in 1998 and followed up thereafter by four assessment reports on human health (2002, 2009, 2015, and 2021).

A review by Abass et al. gives an overview of current trends related to these contaminants and their health effects in humans of the Arctic that is based on data published by AMAP. This reinforced by additional data from the Russian Arctic that is presented Russian scientific literature (Abass et al., 2018). Neurobehavioral, immunological, reproductive, cardiovascular, endocrine and carcinogenic effects of this pollution have also been studied via mother–child cohorts in the Nordic region (see e.g., Weihe & Grandjean, 2021; Weihe et al., 2016).

Fish consumption is the major source of mercury exposure in the Nordic Arctic. However, exposure to legacy contaminants such as mercury and lead is decreasing. This is due, in part, to a dietary transition to more imported food from outside the region. Most of this long-term data related to contaminants within the Arctic have not until recently been readily available within regions. Consequently, it has been difficult to frame appropriate responses and to formulate adequate public policy measures. More knowledge and research are needed about the health impacts of contaminants on human, environmental, and wildlife health if this is to be done (Arctic Monitoring and Assessment Programme (AMAP), 2021; Basu et al., 2022).

The patterns of pollution (POP) can change with climate change. Warming, melting sea ice and glaciers, and permafrost thaw may influence the character and pattern of contaminant exposure, the deposition and cycling of contaminants and toxins between the atmosphere and ocean, and the spread of zoonotic infectious diseases. More research is called for on the influence of climate change on POPs, heavy metals and radionuclides (Arctic Monitoring and Assessment Programme (AMAP), 2002). The growing trend of human migration from smaller communities to larger towns and cities in the Nordic region may be accompanied by changes in diet, contaminant exposure and zoonotic diseases (Arctic Monitoring and Assessment Programme (AMAP), 2021). Furthermore, current waste sites (chemical and mixed domestic wastes) built on permafrost may begin to leak their contents into water systems or onto land masses as increased amounts of permafrost thaw. This may lead to land subsidence (van Oort et al., 2022). Ocean pollution is another important but insufficiently recognized and inadequately controlled component of the current Arctic environment. It poses serious threats to Arctic human health and well-being (Landrigan et al., 2020).

9.6 Zoonoses: Disease Transmission from Animals to Humans

Global pandemics occur as we have seen recently when a new disease suddenly appears against which humans have no immunity. They are often caused by a virus or other pathogens "jumping" from animals to humans. These transfers from animals to humans are called zoonoses and include infections or infectious diseases. Zoonoses are transmitted in many ways, including through direct contact between animals and humans, biting insects, the intake of food and water contaminated with parasites or through the air (de La Rocque et al., 2008). The Spanish flu of 1918, caused by the Influenza A virus found naturally in wild aquatic birds, claimed between 30 and 50 million lives (Taubenberger & Morens, 2006) and is perhaps the best-known and deadliest example. The rabies virus and recent emerging diseases, such as the Ebola, the Zika and the SARS-CoV-2 viruses, are other examples.

Many Northern communities are at least partially dependent upon subsistence activities for their dietary needs and cultural activities. This engages people and wildlife in an intimate relationship that may pose risks for emerging and endemic zoonoses (Reynolds et al., 2022). Environmental stressors are connected in many ways to zoonoses in the Arctic. The frequency of contagious diseases observed in Arctic species has increased in recent years. At higher northern latitudes, warming temperatures can support the survival of invasive tick species and the spread of associated diseases that they carry (Waits et al., 2018). The increased thawing of permafrost may also have the potential of releasing highly resistant spores of other pathogens (Everett, 2020), including viruses revived from ancient permafrost (Alempic et al., 2023). Migratory birds also have the potential to transmit ticks across long distances as well as their antibiotic-resistant genes.

Tularemia from muskrats, anisakis and tapeworm from fish, echinococcosis and rabies from wild or domestic canids, and brucellosis and anthrax from caribou, reindeer, bison or horse, are all relatively common, zoonoses found around the Arctic. Less well understood, potential zoonoses in the region include pathogens such as Erysipelothrix, Leptospira, Chlamydia, Q-fever, Orf virus, a variety of arboviruses viruses transmitted by blood-sucking insects are also in evidence (Reynolds et al., 2022). While we most often focus on zoonotic disease in people, pathogen transmission can occur in the opposite direction, as is thought to have occurred for Giardia in muskoxen (Kutz et al., 2008). Other potentially emerging risks include COVID-19, where spill-over from people has caused widespread outbreaks in farmed mink (Munnink et al., 2020).

Representative examples of these types of zoonoses are presented in Box 9.2 and in review publications (Andersen-Ranberg et al., 2024). The cases highlight critical gaps in monitoring and current knowledge, focusing on environmental stressors and lifestyle factors, and they are examples of current occurrences in the Arctic that inform on critically needed actions to prepare us for the future (Andersen-Ranberg et al., 2024).

Box 9.2 Examples of zoonoses from Reindeer, Fish, Other Species and Climate Change (cc)

- **Examples from reindeer**

The Fennoscandian reindeer population numbers over 660,000 (Riseth et al., 2019) and includes semi-domesticated reindeer, wild tundra reindeer in Norway and forest reindeer in Finland. These animals face increasing anthropogenic and cc challenges and changing infection pressures (Jokelainen et al., 2011; Morten Tryland et al., 2019). As animal husbandry expands into the Nordic Arctic, there are greater risks for diseases to spread from livestock to wild animals. With cc, mosquito-transmitted nematode *Setaria tundra* has been inhabiting more reindeer husbandry area and causes higher frequency of outbreaks of peritonitis in reindeer calves in Finland (Laaksonen, 2010). Climate conditions to the development of Setaria larval infectivity have recently been further focused on Haider et al. (2018). The major limiting factor of the nematode life cycle is low ambient temperature which may prevent development in the mosquito vector, while the mosquito thrives. Toxoplasmosis was found to be associated with the degree of domestication in Fennoscandian semi-domesticated reindeer (Oksanen et al., 1997). The first case of Chronic Wasting Disease (fatal contagious neurodegenerative prion disease in cervids) in Europe was found in a reindeer in south Norway in 2016. The origin of the disease was unknown. Until 2016 with low tests, the prevalence of CWD was unknown (Benestad et al., 2016).

- **Examples from fish**

It is likely that warmer water temperatures will allow pathogens to expand their range, moving further north. The bird intestinal trematode parasite *Cryptocotyle lingua*, an infective parasite, causes "black spot" disease in intermediate fish hosts. In mariculture, particularly for cod, this is a detrimental disease, resulting in spoilage and unsellable fish (Sandneseng, 2006). Another example is *C. lingua*, which can only emerge from its mollusk host when water temperatures are above 10 °C (Tryland et al., 2009). As water temperatures rise, conditions improve for the infective parasite to emerge and have greater opportunities to expand its range. The salmonid parasite, *Tetracapsuloides bryosalmonae* is limited by temperature. While widespread in freshwater invertebrate bryozoans ("moss animals") and salmonids, *T. bryosalmonae* causes proliferative kidney disease when water temperatures remain above 14 °C for enough long periods of time. *T. bryosalmonae* is responsible for severe disease and population decline in parr (young Salmoninae spp.) in at least two Norwegian rivers (Tryland et al., 2009). Warmer waters are also a factor in increasing rates of bacterial furunculosis and protozoan-caused ichthyophonosis and whirling diseases. This supports the free-living stages of fish pathogens and possibly stressing the fish and leave them more susceptible to diseases (Bradley et al., 2005). Zoonotic fish parasites *Diphyllobothrium spp., A. simplex*, and *P. decipiens* are likely to have faster hatching and development of larvae, increased abundance in hosts because of cc.

- **Examples from other species**

The world is changing and parasites adapt. Increasing temperatures are predicted to expand the range and abundance of parasites *U. pallikuukensis, P. odocoilei, and E. rangiferi* larvae. High environmental temperature reduces the Trichinella larva survival in carrions. In Finland, *T. nativa* appears to thrive better than ever before (Oksanen et al., 2018). The Nordic countries face new parasitological challenges due to changes in populations of parasites and their hosts and the spread of new parasites to the region due to climate change (Deksne et al., 2020). *T. gondii, E. multilocularis, E. granulosus*, pathogenic emerging zoonotic diseases in the Nordic countries, have been considered likely to have better opportunities for their sexual life cycle and to be spread to more hosts as a result of cc (Davidson et al., 2011; Wahlström et al., 2011).

9.7 Species on the Move

We are presently witnessing one of the largest climate-driven global redistribution of species since the last glacial maximum. While the traditional geographic range of species can fluctuate dynamically in space and time, recent climate changes can cause a major migration of species populations. If "cold species" cannot adapt to warming local conditions, they will seek new habitat towards the poles, higher up the mountains, or deeper down in the oceans (Pecl et al., 2017, 2023). Climatic changes can lead to shifts in the geographic boundaries of pathogens, hosts, and vectors. This can amplify the transmission of endemic climate-sensitive pathogens (Kovats et al., 2001; Kutz et al., 2005; Parkinson et al., 2014). Climate change can transform the geographic distribution and seasonal patterns of transmission for a range of infectious and parasitic diseases and push these northward. Likewise, Arctic-adapted species may face extinction at the top of a warming world. Both human-specific and zoonotic pathogens can cause disease in Arctic human populations. Non-zoonotic pathogens can also cause significant economic damage to the region, threatening trade, livestock production, and the harvesting and conservation of wildlife (Evengård & Thierfelder, 2021).

Predicting the impact of climate change on infectious diseases in the Nordic Arctic species is a difficult task. Scientific research can be limited by large gaps in baseline knowledge and documentation, difficulties in logistics, and the higher costs of investigation within the region. Likewise, disease trends, even in a stable ecosystem, are difficult to predict because of the complexity of the host—agent—environment relationships. While specific disease patterns are hard to predict, several factors related to climate change that may influence infectious diseases now and, in the future, can be identified. These include increasing environmental temperatures, changes in biodiversity, introduction of domestic animals and invasive species, and an increase in the level of persistent environmental contaminants (Bradley et al., 2005).

Changes in temperature, precipitation and humidity affect the rate of development, survival, reproduction and migration of pathogens and disease vectors. They influence the spread of vector-borne infections and zoonotic diseases, including to new hosts (van Oort et al., 2022). Higher temperatures can allow infected carriers to survive the unfavorable periods of the year, increasing their population size and range of the species. It also increases the opportunity for transmission of infection to humans. In general, the Arctic with its low diversity of animal, plant and microbial species, and increasing surface temperature will experience more severe changes in infectious disease patterns than other regions of the globe (Hassol, 2004; Miller & Ruiz, 2014).

Climate change also supports the spread of microorganisms in many other ways. For thousands of years, as humans have migrated across land and explored the oceans, they have carried species with them. Around 480,000 invasive species are now estimated to have been introduced around the world by humans (Pimentel et al., 2001). Some of these introductions have come to the Arctic and to the Nordic portion of it. More are likely to An ice-free Northern Ocean Route between Europe and the Northern Hemisphere (Miller & Ruiz, 2014).

It is also probable that microorganisms, including pathogens like anthrax, will re-appear in ecosystems due to thawing of the burial places of people and animals in permafrost and glaciers within the Arctic. The climate-change thawing of landscapes can also bring about changing land-use practices that can have harmful health consequences. As an example, the increasing accessibility of precious metals across the far North is likely to initiate mining projects that have the potential to expose buried microbes that may be revived. It is therefore highly recommended that studies of potential new mining sites be undertaken ahead of any resource exploitation to identify and avoid any such health threats.

9.8 The Need for Long-Term Data Collection

In order to adequately address these and other types of health challenges within the Arctic and to formulate adequate policy responses within the Nordic area and other sectors, it is important that long-term data collection be undertaken. Such data needs to be secured in several areas regarding the environment, humans and wildlife. This data, and its analysis, also needs to be made widely available to researchers, government decision makers and members of the public–especially the residents of the North. To be of consequence, this data collection effort must also be undertaken across borders. This requires both cooperative research efforts among Nordic teams of researchers as well as collaborative efforts with other circumpolar countries and European partners. The Arctic ecosystems do not recognize national borders and consequently data collection must be undertaken in a similarly inclusive manner. Both Russian and North American scientists need to be part of the effort. This can be difficult give the current geopolitical conditions described in Chap. 7 of this volume, but an effort must be made. Among themselves the Nordics have a long-established record of such undertakings (see Box 9.3). But even here there exists some inner diversity among them regarding what type of data they collect, how they present this information, and how they make it available to other researchers, policymakers and the public.

Box 9.3: An example of monitoring wildlife data in Scandinavia

Monitoring of wildlife diseases started in Scandinavia as sporadic postmortem examinations in the early 20th century. In 1945 a monitoring program for wildlife health was initiated. The program is today an integrated part of the National Environmental Monitoring programs in Sweden. The total material today comprises more than 80,000 recorded investigations. There are similar programs in Denmark, Finland, and Norway, but no comparable program exists in other Northern European countries. The program has led to discoveries of new diseases in many mammals and birds, and demonstration of several environmental pollutants, such as mercury, lead, and cadmium, in wildlife. The success of the program is due to several factors such as: a rich wildlife with economically important game species; location of the program at the National Veterinary Institute with excellent facilities and competent staff, good cooperation with hunter and conservation organizations, stable financial support, a cool climate, and a successful relationship with media. Current work also includes more focused investigations on different specific pathogens and pollutants, as well as investigations in single animal species (Morner 2002). The annually published animal disease surveillance reports may serve as a model for other arctic and subarctic regions.

9.9 Data Procurement via Citizen Science (CS), Community-Based Monitoring (CBM) and Participatory Research in the Nordic Context

Despite the need for a One Health type of response to climate change in the Nordic Arctic, conducting actual environmental monitoring such regions can often be difficult (Mosites et al., 2018). This is due to various organizational, bureaucratic, and economic hurdles that need to be overcome. One partial solution to these problems is to include more northern residents in the undertaking. These individuals usually maintain a relationship with the environment and climate change has a major impact on their communities (Mosites et al., 2018). Important signs of changing ecosystems are noticed relatively early by individuals living in close contact with nature. The involvement of such citizens in integrated science design and implementation should be encouraged. Researchers in the Nordic countries have demonstrated an increased interest in exploring various options, testing methods, and in support of Citizen Science (CS). Such community-based observation and monitoring efforts can help to supply both needed data and a mechanism for information sharing within affected communities and assist broader efforts at public policy formulation decision-making.

The Local Environmental Observer (LEO) Network is a good example of data collection via CS in the Nordic Arctic. It connects local community observers with scientists, government agencies and health care providers. LEO Network members share data and unusual environmental events with their counterparts from outside the area that allows for an enhanced understanding of the changing world. Some examples include the reporting of abnormal behavior patterns on the part of reindeer in Finnish Lapland in 2019; an early spring migration of whooper swans in Finland in 2020; and the sudden and unusual appearance of tuna in northern Norway in 2021 (LEO network, 2023).

A large international project (CLINF) based in the Nordic North organized a citizen-based surveillance in 2018 (see Fig. 9.1). This undertaking identified some 4500 ticks found on animals or humans in the northern half of Sweden. Its sampling effort produced 35 specimens of adult Hyalomma (Grandi et al., 2020). These are large ticks (20 mm after feeding), that are normally associated with the Mediterranean area. Hyalomma had previously been found in Sweden as nymphs on migratory birds and, in one case, as an adult an imported horse. However, such large numbers of adult ticks on non-imported animals have never been observed. These ticks can act as vectors for a range of pathogens and may not have been identified without the participation of such citizen-based science. These efforts have the potential of providing important information and input to the subsequent policy formulation affecting the region in the areas of the environment and health. Additional study of their operation and impact across the Nordic region seems warranted.

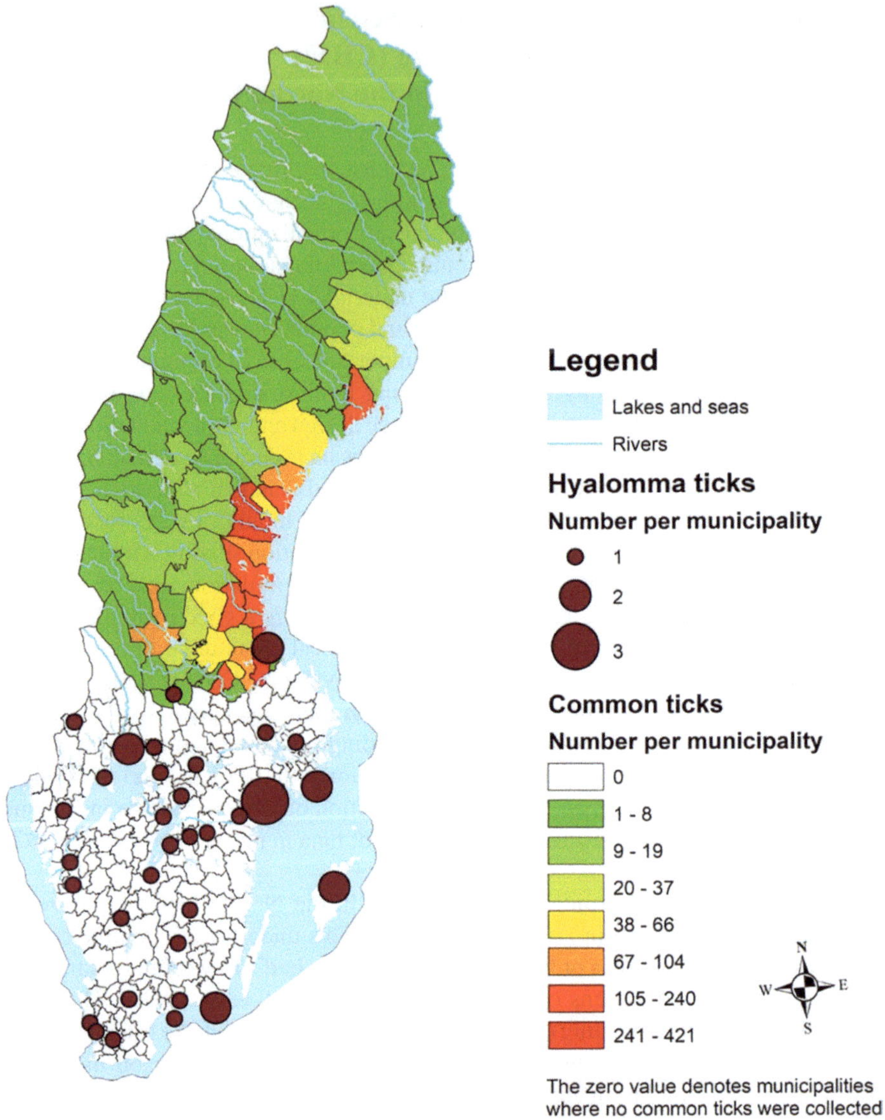

Fig. 9.1 CLINF ticks survey conducted via a campaign of citizen science in 2018. As a result, the first ever observation of adult Hyalomma ticks was made in Sweden. Data collected by SVA, cartographer: Tomas Thierfelder, SLU-ET

9.10 Conclusions

Throughout human history, various communities have experienced epidemics. The challenge of living in nature surrounded by microorganisms is a delicate balance. New diseases can appear when ecosystems change, and humans come in closer contact with new pathogens. As climate change continues new species and new diseases are moving into the northern regions of the Nordic area at a speed not previously recorded. An improved international system of study and surveillance of the relationship between climate change and infectious diseases is much needed.

We have discussed how contaminants and zoonoses can have an impact on life in the Nordic region. We have also identified the need for a well-functioning system for international collaboration in surveillance and data collection. We have pointed to the need for the involvement of local communities in such efforts. Microorganisms and species do not respect human borders. To achieve optimal control of infectious diseases in the Arctic, we need trustful international collaboration, action and effective harmonization of the data which is collected. The adoption of a One Health perspective can be of help in this regard. We must also find new means to involve national, regional and international decision makers in the effort to formulate effective policy responses to the environmental and health challenges we have identified. If we fail in doing so, there might be real chaos in the future.

References

Abass, K., Emelyanova, A., & Rautio, A. (2018). Temporal trends of contaminants in arctic human populations. *Environmental Science and Pollution Research International, 25*(29), 28834–28850. https://doi.org/10.1007/s11356-018-2936-8

Alempic, J.-M., Lartigue, A., Goncharov, A. E., Grosse, G., Strauss, J., Tikhonov, A. N., Fedorov, A. N., et al. (2023). An update on eukaryotic viruses revived from ancient permafrost. *Viruses, 15*(2), 564. https://doi.org/10.3390/v15020564

Andersen-Ranberg, E., Nymo, I. H., Jokelainen, P., Emelyanova, A., Jore, S., Laird, B., Davidson, R. K., et al. (2024). Environmental stressors and zoonoses in the arctic: Learning from the past to prepare for the future. *The Science of the Total Environment, 957*(December), Article 176869. https://doi.org/10.1016/j.scitotenv.2024.176869

Arctic Council. (2023). Addressing Arctic Pollution. Arctic Council. 2023. https://arctic-council.org/explore/topics/pollutants/

Arctic Monitoring and Assessment Programme (AMAP). (2002). *AMAP Assessment 2002: Human Health in the Arctic*. https://www.amap.no/documents/doc/amap-assessment-2002-human-health-in-the-arctic/95

Arctic Monitoring and Assessment Programme (AMAP). (2017). *Snow, Water, Ice and Permafrost in the Arctic (SWIPA) 2017*. https://oaarchive.arctic-council.org/handle/11374/2105

Arctic Monitoring and Assessment Programme (AMAP). (2021). *AMAP Assessment 2021: Human Health in the Arctic*. Tromsø, Norway. https://www.amap.no/documents/doc/amap-assessment-2021-human-health-in-the-arctic/3593

Basu, N., Abass, K., Dietz, R., Krümmel, E., Rautio, A., & Weihe, P. (2022). The Impact of Mercury Contamination on Human Health in the Arctic: A State of the Science Review. *The Science of the Total Environment, 831*(July), Article 154793. https://doi.org/10.1016/j.scitotenv.2022.154793

Benestad, S. L., Mitchell, G., Simmons, M., Ytrehus, B., & Vikøren, T. (2016). First case of chronic wasting disease in Europe in a Norwegian free-ranging Reindeer. *Veterinary Research, 47*(1), 88. https://doi.org/10.1186/s13567-016-0375-4

Berner, J. S., Jore, K. A., & Rautio, A. (2024). One health in the Arctic—Connections and actions. *International Journal of Circumpolar Health, 83*(1), 2361544. https://doi.org/10.1080/22423982.2024.2361544

Bokhorst, S., Højlund Pedersen, S., Brucker, L., Anisimov, O., Bjerke, J. W., Brown, R. D., Ehrich, D., et al. (2016). Changing arctic snow cover: A review of recent developments and assessment of future needs for observations, modelling, and impacts. *Ambio, 45*(5), 516–537. https://doi.org/10.1007/s13280-016-0770-0

Bradley, M., Kutz, S. J., Jenkins, E., & O'Hara, T. M. (2005). The potential impact of climate change on infectious diseases of Arctic Fauna. *International Journal of Circumpolar Health, 64*(5), 468–477. https://doi.org/10.3402/ijch.v64i5.18028

de Castañeda, R. R., Villers, J., Faerron Guzmán, C. A., Eslanloo, T., de Paula, N., Machalaba, C., Zinsstag, J., Utzinger, J., Flahault, A., & Bolon, I. (2023). One health and planetary health research: Leveraging differences to grow together. *The Lancet Planetary Health, 7*(2), e109–e111. https://doi.org/10.1016/S2542-5196(23)00002-5

Dai, H. (2021). Roles of surface Albedo, surface temperature and carbon dioxide in the seasonal variation of arctic amplification. *Geophysical Research Letters, 48*(4): e2020GL090301. https://doi.org/10.1029/2020GL090301

Davidson, R., Simard, M., Kutz, S. J., Kapel, C. M. O., Hamnes, I. S., & Robertson, L. J. (2011). Arctic parasitology: Why should we care? *Trends in Parasitology, 27*(6), 239–245. https://doi.org/10.1016/j.pt.2011.02.001

De La Rocque, S., Caya, F., El Idrissi, A. H., Mumford, L., Belot, G., Carron, M., Sreedharan, R., et al. (2019). One health operations: A critical component in the international health regulations monitoring and evaluation framework. *Revue Scientifique Et Technique (International Office of Epizootics), 38*(1): 303–314. https://doi.org/10.20506/rst.38.1.2962

Deksne, G., Davidson, R. K., Buchmann, K., Kärssin, A., Kirjušina, M., Gavarāne, I., Miller, A. L., et al. (2020). Parasites in the changing world—Ten timely examples from the Nordic-Baltic region. *Parasite Epidemiology and Control, 10*(August), Article e00150. https://doi.org/10.1016/j.parepi.2020.e00150

Evengård, B., Destouni, G., Kalantari Z., Albihn, A., Björkman, C., Bylund, H., Jenkins, E., et al. (2021). Healthy ecosystems for human and animal health: Science diplomacy for responsible development in the Arctic: The Nordic Centre of excellence, Clinf.Org (climate-change effects on the epidemiology of infectious diseases and the impacts on Northern Societies). *Polar Record, 57*, e39. https://doi.org/10.1017/S0032247421000589

Evengård, B., & Thierfelder, T. (2021). CLINF: Climate-change effects on the epidemiology of infectious diseases, and the associated impacts on Northern Societies. In D.C. Nord (Ed.), *Nordic perspectives on the responsible development of the Arctic: Pathways to action* (pp.49–70). Springer Polar Sciences. Springer International Publishing. https://doi.org/10.1007/978-3-030-52324-4_3

Everett, L., (Ed.). (2020). *Understanding and Responding to Global Health Security Risks from Microbial Threats in the Arctic: Proceedings of a Workshop.* National Academies Press. https://doi.org/10.17226/25887

Gartler, S., Scheer, J., Meyer, A., Abass, K., Bartsch, A., Doloisio, N., Falardeau, J., et al. (2025). A transdisciplinary, comparative analysis reveals key risks from arctic permafrost thaw. *Communications Earth & Environment, 6*(1), 21. https://doi.org/10.1038/s43247-024-01883-w

Grandi, G., Chitimia-Dobler, L., Choklikitumnuey, P., Strube, C., Springer, A., Albihn, A., Jaenson, T. G. T., & Omazic, A. (2020). First records of adult Hyalomma Marginatum and H. Rufipes Ticks (Acari: Ixodidae) in Sweden. *Ticks and Tick-Borne Diseases, 11*(3), 101403. https://doi.org/10.1016/j.ttbdis.2020.101403

Haider, N., Laaksonen, S., Kjær, L. J., Oksanen, A., & Bødker, R. (2018). The annual, temporal and spatial pattern of Setaria Tundra outbreaks in Finnish Reindeer: A mechanistic transmission model approach. *Parasites & Vectors, 11*(1), 565. https://doi.org/10.1186/s13071-018-3159-z

Hassol, S. J. (2004). *Impacts of a warming climate—Arctic climate impact assessment*. Cambridge University Press.

Heininen, L., Everett, K., Padrtova, B., & Reissell, A. (2020). Arctic policies and strategies—Analysis, synthesis, and trends. Report. International Institute for Applied Systems Analysis. https://doi.org/10.22022/AFI/11-2019.16175

Hovelsrud, G. K., Poppel, B., van Oort, B., & Reist, J. D. (2011). Arctic societies, cultures, and peoples in a changing cryosphere. *Ambio, 40*(1), 100–110. https://doi.org/10.1007/s13280-011-0219-4

Jahn, A., Kay, J. E., Holland, M. M., & Hall, D. M. (2016). How predictable is the timing of a summer ice-free arctic? *Geophysical Research Letters, 43*(17), 9113–9120. https://doi.org/10.1002/2016GL070067

Jokelainen, P., Isomursu, M., Näreaho, A., & Oksanen, A. (2011). Natural Toxoplasma Gondii infections in European Brown Hares and Mountain Hares in Finland: proportional mortality rate, antibody prevalence, and genetic characterization. *Journal of Wildlife Diseases, 47*(1), 154–163. https://doi.org/10.7589/0090-3558-47.1.154

Kovats, R. S., Campbell-Lendrum, D. H., McMichael, A. J., Woodward, A., & Cox, J. S. (2001). Early effects of climate change: Do they include changes in vector-borne disease? Philosophical Transactions of the Royal Society of London. *Series b, Biological Sciences, 356*(1411), 1057–1068. https://doi.org/10.1098/rstb2001.0894

Kutz, S. J., Hoberg, E. P., Polley, L., & Jenkins, E. J. (2005). Global warming is changing the dynamics of arctic host-parasite systems. *Proceedings of the Royal Society b: Biological Sciences, 272*(1581), 2571–2576. https://doi.org/10.1098/rspb.2005.3285

Kutz, S. J., Andrew Thompson, R. C., Polley, L., Kandola, K., Nagy, J., Wielinga, C. M., & Elkin, B. T. (2008). Giardia Assemblage A: Human genotype in Muskoxen in the Canadian Arctic. *Parasites & Vectors, 1*(1), 32. https://doi.org/10.1186/1756-3305-1-32

Kutz, S., & Tomaselli, M. (2019). "Two-Eyed Seeing' supports wildlife health. *Science, 364*(6446), 1135–1137. https://doi.org/10.1126/science.aau6170

de La Rocque, S., Rioux, J. A., & Slingenbergh, J. (2008). Climate change: Effects on animal disease systems and implications for surveillance and control. *Revue Scientifique Et Technique (International Office of Epizootics), 27*(2), 339–354.

Laaksonen, S. (2010). Setaria Tundra, an Emerging Parasite of Reindeer, and an Outbreak It Caused in Finland in 2003–2006, February. https://helda.helsinki.fi/handle/10138/18996

Landrigan, P. J., Stegeman, J. J., Fleming, L. E., Allemand, D., Anderson, D. M., Backer, L. C., Brucker-Davis, F., et al. (2020). *Human Health and Ocean Pollution, 86*(1), 151. https://doi.org/10.5334/aogh.2831

Larsen, J. N., & Fondahl, G. (2015). *Arctic human development report: Regional processes and global linkages*. Nordic Council of Ministers. http://norden.diva-portal.org/smash/record.jsf?pid=diva2:788965

LEO network. (2023). Local Environmental Observer (LEO) Network. www.Leonetwork.Org. https://www.leonetwork.org

Miller, A. W., & Ruiz, G. M. (2014). *Arctic shipping and marine invaders*. https://doi.org/10.1038/nclimate2244

Morner, T. (2002). Health monitoring and conservation of wildlife in Sweden and Northern Europe. *Annals of the New York Academy of Sciences, 969*(October), 34–38. https://doi.org/10.1111/j.1749-6632.2002.tb04347.x

Mosites, E., Lujan, E., Brook, M., Brubaker, M., Roehl, D., Tcheripanoff, M., & Hennessy, T. (2018). Environmental observation, social media, and one health action: A description of the local environmental observer (LEO) network. *One Health, 6*, 29–33. https://doi.org/10.1016/j.onehlt.2018.10.002

Munnink, B. B. O., Sikkema, R. S., Nieuwenhuijse, D. F., Jan Molenaar, R., Munger, E., Molenkamp, R., van der Spek, A., et al. (2020). Jumping Back and Forth: Anthropozoonotic and Zoonotic Transmission of SARS-CoV-2 on Mink Farms. bioRxiv. https://doi.org/10.1101/2020.09.01.277152

Oksanen, A., Åsbakk, K., Nieminen, M., Norberg, H., & Näreaho, A. (1997). Antibodies against toxoplasma Gondii in Fennoscandian Reindeer—Association with the degree of domestication. *Parasitology International, 46*(4), 255–261. https://doi.org/10.1016/S1383-5769(97)00033-0

Oksanen, A., Interisano, M., Isomursu, M., Heikkinen, P., Tonanzi, D., Oivanen, L., & Pozio, E. (2018). Trichinella Spiralis prevalence among wildlife of a boreal region rapidly reduced in the absence of spillover from the domestic cycle. *Veterinary Parasitology, 262*(October), 1–5. https://doi.org/10.1016/j.vetpar.2018.09.002

Parkinson, A. J., Evengard, B., Semenza, J. C., Ogden, N., Børresen, M. L., Berner, J., Brubaker, M., et al. (2014). Climate change and infectious diseases in the arctic: Establishment of a circumpolar working group. *International Journal of Circumpolar Health, 73*, 25163. https://doi.org/10.3402/ijch.v73.25163

Pecl, G. T., Araújo, M. B., Bell, J. D., Blanchard, J., Bonebrake, T. C., Chen, I.-C., Clark, T. D., et al. (2017). Biodiversity redistribution under climate change: Impacts on ecosystems and human well-being. *Science, 355*(6332), eaai9214. https://doi.org/10.1126/science.aai9214

Pecl, G. T., Kelly, R., Lucas, C., van Putten, I., Badhe, R., Champion, C., I-Ching Chen, et al. (2023). Climate-Driven 'Species-on-the-Move' Provide Tangible Anchors to Engage the Public on Climate Change. People and Nature n/a (n/a). https://doi.org/10.1002/pan3.10495

Pimentel, D., McNair, S., Janecka, J., Wightman, J., Simmonds, C., O'Connell, C., Wong, E., et al. (2001). Economic and environmental threats of Alien Plant, animal, and microbe invasions. *Agriculture, Ecosystems & Environment, 84*(1), 1–20. https://doi.org/10.1016/S0167-8809(00)00178-X

Rantanen, M., Yu Karpechko, A., Lipponen, A., Nordling, K., Hyvärinen, O., Ruosteenoja, K., Vihma, T., & Laaksonen, A. (2022). The arctic has warmed nearly four times faster than the globe since 1979. *Communications Earth & Environment, 3*(1), 1–10. https://doi.org/10.1038/s43247-022-00498-3

Reynolds, A., Kutz, S., & Baker, T. (2022). A holistic approach to one health in the arctic. In M. Tryland (Ed.), *Arctic one health: Challenges for northern animals and people* (pp. 21–45). Springer International Publishing. https://doi.org/10.1007/978-3-030-87853-5_2

Riseth, J. Å., Tømmervik, H., & Forbes, B. C. (2019). Sustainable and resilient reindeer herding. In M. Tryland & S. Kutz (Eds.), *Reindeer and Caribou* (pp. 23–43). CRC Press.

Samenow, J. (2021). North pole surges above freezing in the dead of winter, stunning scientists. Washington Post, December 3, 2021. https://www.washingtonpost.com/news/capital-weather-gang/wp/2018/02/26/north-pole-surges-above-freezing-in-the-dead-of-winter-stunning-scientists/

Sandneseng, E. (2006). Variation in abundance, diet, otolith zone patterns and black spot disease (*Cryptocotyle lingua*) of 0-Group Coastal Cod (*Gadus morhua* L.) in Northern Norway', December. https://munin.uit.no/handle/10037/430

Schiermeier, Q. (2018). Droughts, heatwaves and floods: How to tell when climate change is to blame. *Nature, 560*(7716), 20–22. https://doi.org/10.1038/d41586-018-05849-9

Taubenberger, J. K., & Morens, D. M. (2006). 1918 Influenza: The mother of all pandemics. *Emerging Infectious Diseases, 12*(1), 15–22. https://doi.org/10.3201/eid1201.050979

Tryland, M., Godfroid, J., & Arneberg, P. (2009). Impact of climate change on infectious diseases of animals in the Norwegian Arctic. Kortrapport/Brief Report Series. The Norwegian Polar Institute. https://www.semanticscholar.org/paper/Impact-of-climate-change-on-infectious-diseases-of-Tryland-Godfroid/de539e41beec028924416cb0f8ffc8d65b17fb9a

Tryland, M., Nymo, I. H., Romano, J. S., Mørk, T., Klein, J., & Rockström, U. (2019). Infectious disease outbreak associated with supplementary feeding of semi-domesticated reindeer. *Frontiers in Veterinary Science, 6*. https://www.frontiersin.org/articles/10.3389/fvets.2019.00126

United Nations. (1987). Brundtland Report of the World Commission on Environment and Development: Our Common Future. 43. United Nations. https://www.are.admin.ch/are/en/home/medien-und-publikationen/publikationen/nachhaltige-entwicklung/brundtland-report.html

University of Jyväskylä. (2023). Course: Introduction to Planetary Well-Being—BENY2021 (PW MOOC I). JYU Online Courses. https://onlinecourses.jyu.fi/course/view.php?id=29

van Oort, B., Tronstad Lund, M., & Brisebois, A. (2022). Climate change in Northern Regions. In M. Tryland (Ed.), *Arctic one health: Challenges for northern animals and people* (pp. 79–119). Springer International Publishing. https://doi.org/10.1007/978-3-030-87853-5_4

Waits, A., Emelyanova, A., Oksanen, A., Abass, K., & Rautio, A. (2018). Human infectious diseases and the changing climate in the arctic. *Environment International, 121*(Pt 1), 703–713. https://doi.org/10.1016/j.envint.2018.09.042

Wahlström, H., Isomursu, M., Hallgren, G., Christensson, D., Cedersmyg, M., Wallensten, A., Hjertqvist, M., Davidson, R. K., Uhlhorn, H., & Hopp, P. (2011). Combining information from surveys of several species to estimate the probability of freedom from Echinococcus Multilocularis in Sweden, Finland and Mainland Norway. *Acta Veterinaria Scandinavica, 53*(1), 9. https://doi.org/10.1186/1751-0147-53-9

Weihe, P., Debes, F., Halling, J., Petersen, M. S., Muckle, G., Odland, J. O., Dudarev, A., et al. (2016). Health effects associated with measured levels of contaminants in the arctic. *International Journal of Circumpolar Health, 75*(1), 1–12. https://doi.org/10.3402/ijch.v75.33805

Weihe, P. M., & Grandjean, P. (2021). Cohort studies of Faroese children concerning potential adverse health effects after the mothers' exposure to marine contaminants during pregnancy. *Acta Veterinaria Scandinavica, 24* ((Suppl 1)), S7.

Birgitta Evengård is a specialist in infectious diseases and clinical immunology and has worked as a senior consultant besides performing clinical research and teaching at Karolinska Institutet, at Stockholm University (malaria immunology), Pasteur Institute, Lille (schistosomiasis), Harvard Medical School (filariasis), University of California Irvine (Borna disease virus and Chronic fatigue syndrome), Columbia University (climate change and infectious diseases). She received her Ph.D. in 1989 from the Karolinska Institutet, and was professor in Clinical Parasitology at the Karolinska Institute in 2006 and the first female professor with a chair in Infectious Diseases in Sweden at Umeå University in 2007. Since 2007 the focus for her research has been on climate change and transition of infectious diseases in the North. In 2016 she received funding as a lead from the Nordic Council of Ministers for a Nordic Centre of Excellence, a six-year project involving all Nordic countries, England, and Russia, www.clinf.org. 2018 Evengård was awarded the Jack Hildes Award for her achievements in Arctic health. She has been a member of the Lancet Commission for Health in the Arctic, an Executive Board member of EU Polar-Net 1 and 2, and a chair of the AMAP/SDWG Health Group in the Arctic Council. She is the author of 11 books, 23 book chapters and more than 160 publications in peer-reviewed journals, with more than 12,000 citations, h-index 39. Evengård has promoted gender equality throughout her career in academia and has performed groundbreaking work with gender equality in clinical care. In 2023, both national and international media identified her as one of the most visible researchers working on climate change at Umeå University.

Anastasia Emelyanova works in the Arctic Health research group of the University of Oulu. She is a Vice-Lead of the UArctic Thematic Network on Health and Well-being. Her research interests include One Health, healthy ageing in the Arctic communities, demography and population changes across the Arctic region, northern health and well-being, and policy in these respective fields. She holds academic degrees in Arctic Health from the University of Oulu (Finland) and Northern Arctic Federal University (Russia). In 2024, she received the Jens Peder Hart Hansen Award that recognizes excellent emerging researchers in the field of circumpolar health. She coordinates the project "The Nordic Arctic Healthy Aging Communities: Safe and Inclusive Outdoor

Environment and Public Spaces" (2023–2026) as well as contribute to several ongoing international projects related to circumarctic health such as ArcSolution (2024–2028), One Health in Northern Communities and Ecosystems (2020–2025) among others.

Chapter 10
(Ab)using the Land: Environmental and Climate Justice in Public Policy of the Nordic North

Corine Wood-Donnelly, Romain Chuffart, Olof Stjernström, Keith Larson, and Håkan T. Sandersen

Abstract This chapter discusses the context and challenges for environmental and climate justice in the northern or Arctic parts of the Nordic region. These spaces present interesting opportunities to examine notions of exceptionalism and inner diversity with regard to Nordic policies and practices concerning land use. In this chapter, we ask: "Who can do what with the land?" and "How are these activities shaped by, and shaping, environmental and climate change?" To answer these questions, we examine historical and contemporary land use dynamics vis-à-vis colonialism, governance, and rights and assess what tensions may exist between state regulations, Indigenous rights, local communities' interests, and environmental climate imperatives. In our discussion, we use the several lenses of justice including the distribution of harms and benefits, access to participation in decision-making processes and recognition. In so doing we show how public and private land governance frameworks affect local communities' interests and the implementation of the rights of Indigenous peoples in the Nordic countries relative to climate and the environment.

Keywords Nordic land use policy · Environmental justice · Climate justice · Indigenous rights · Indigenous peoples

---

C. Wood-Donnelly (✉) · O. Stjernström · H. T. Sandersen
Nord University, Bodø, Norway
e-mail: c.wood-donnelly@nord.no

R. Chuffart
University of Akureyri, Akureyri, Iceland

K. Larson
Umeå University, Umeå, Sweden

D. C. Nord (ed.), *Contemporary Nordic Public Policies*, Springer Polar Sciences,
https://doi.org/10.1007/978-3-032-11292-7_10

10.1 Introduction

From the perspective of an everyday schoolroom globe, the Nordic countries are seen as being northern, sharing the Arctic with other high-latitude countries. Even the geographic labels for these countries, such as *Norden* or *Nordic*, emerge etymologically from Old Norse with implications of both a cardinal direction and coming from a location. The Latin word *nordicus,* meaning 'of the north,' is probably a Latinization of the older Germanic/Scandinavian word, reinforcing the understanding that these countries lay north of the rest of Europe. This northernness carries other cultural associations including the mythological understanding of the north wind shaping seasons and landscapes, along with the people of the North. It is also the most recent area of the Earth to be exposed by the retreating ice sheet at the end of the last ice age.

These circumstances of the geography of the Nordic region have also shaped its social, economic, and political development in relative geographic isolation at the periphery of Europe. This peripheral position created a unique cultural ecosystem that has contributed to the development of strong welfare states, the prioritisation of broad social interest and a distinctive approach to sustainable development through stakeholder collaboration (Strand, 2024). These physical conditions accompany a history of peaceful development and a transformation from old kingdoms to modern democracies. Within this landscape, we also find Nordic philosophical and cultural practices connected to the environment through concepts such as *friluftsliv* (open-air life or outdoor recreation) and *allemansrätten* (freedom to roam or right of public access). Both include an emphasis on individual responsibility for connecting with and caring for nature. It is from within this cultural context that philosophers such as Arne Næss developed his eight principles for Deep Ecology (Naess, 1989) and Georg Henrik von Wright books like "Humanismen som livshållning" (Humanism as a way of life) (von Wright, 1978).

Although the Nordics form Europe's northern periphery, they also contain an even more remote periphery known as the Arctic, *Nordområder*, *Lapland*, or *Norðrið*. The Nordic region's overall reputation for social and environmental responsibility stems largely from the post-1945 era. However, it is challenged in these northern areas, where colonial histories, Indigenous marginalization, and resource exploitation have long created tensions between the views and preferences of national capitals (Helsinki, Stockholm, Copenhagen, Oslo, and Reykjavik) and those of Indigenous and other local communities. These tensions shape ongoing debates over territorial rights, the legitimacy of resource extraction, and policy choices tied to the EU Green Deal and national interests. Historical legacies also reinforce path-dependencies and socio-economic frameworks that often sideline the periphery's concerns.

Public policy regarding land use in the Nordic North involves overlapping claims in which climate and environmental challenges intersect. Although sometimes the North is seen as an "exceptional" governance space, we nonetheless seek to identify and explain specific established patterns of policy development related to land use that can be seen both there and elsewhere on the globe. We seek to link the need for land use policy innovation to climate and environmental justice demands. In so doing, we ask two questions: "Who can do what with the land?" and "How are these activities shaped by the environment and climate change?" In providing answers, we examine the political and economic forces influencing land use decisions within and outside the Nordic North.

This chapter first considers environmental and climate history through perspectives of justice, considering distributional, procedural and recognitional aspects of justice theory. Second, we examine land use and Nordic colonialism by reflecting on how past power imbalances shape current resource policies. Third, we discuss land use rights vis-à-vis the state and how these are negotiated between national priorities and local and Indigenous claims. Finally, the chapter looks at contemporary land use policies and their variations across the Nordic countries. The chapter suggests that rethinking current frameworks through inclusive and justice-oriented approaches can foster more balanced and equitable regional outcomes.

10.2 Environmental (Justice) History and Climate (Justice) History

Environmental and climate justice provides a useful way to examine land use policies in the Nordic region, particularly regarding their effects on local and decolonial politics and managerial policies that affect Sámi peoples and local communities. Although Nordic states are often seen as global leaders in environmental policy (Reyes, 2021), they have not always recognised the rights and perspectives of the Sámi people and other affected groups. Indeed, they continue to fall short in some aspects of climate protection (McCauley et al., 2022).

As we examine the gap between convergence and divergence of policy and its effects on the determination of land use in the high north of the Nordics, the concepts of environmental and climate justice give perspective in framing this landscape. They can help us understand why the northern realms of the Nordics are sometimes conceived of as a "sacrifice zone" (Skorstad, 2022). They also assist in explaining why the state is sometimes taken to court over energy development (Norwegian Human Rights Institution, 2023), decisions about hunting rights (Högsta Domstolen, 2020), and the allocation of fishing quotas (Tómas, 2023).

Environmental justice concerns how environmental harm and benefits are shared across different groups and places (Wood-Donnelly, 2023a). It considers local and global impacts, recognising that decisions made in one area can have consequences far beyond it. A fair environmental system, in this view, is one where the negative effects of pollution, resource extraction, and land degradation do not fall disproportionately on some groups. It is also one in which other distant communities do not benefit from cleaner environments and economic opportunities at the expense of certain other locales, including the North.

Some approaches to climate justice focuses on the sometimes-unequal effects of climate change and the responsibility of governments and organisations to address these. For example, Skillington (2023) argues that climate justice is about more than reducing emissions or limiting damage; it is also about ensuring that policies consider the needs and rights of those most affected. Indigenous peoples, often depend on stable ecosystems for their ways of life and environmental rights. In this context, implementing justice-based climate policies can also mean understanding, integrating and considering the rights of land users suffering most from climate change effects.

Although both environmental and climate justice provide a conceptual starting point to assess the impacts of how land is managed and used in the Nordics, they remain quite abstract. They need to be supplemented with practical understandings of justice theory. In this regard, three aspects of justice: *distribution*, *process*, and *recognition,* become useful to explain and assess how past and present Nordic decisions and policies have shaped the dynamics of land use and the resulting degrees of exceptionalism found in their provisions and impacts. Beyond this, these more concrete aspects of *doing* justice in practice serve as a basis for understanding unequal governance within each Nordic state.

The first aspect of this perspective is understanding the distribution of benefits and harms within the local environment and the effects of climate change. This is referred to as distributive justice. In the Nordic North, heavy extractive industries such as mining, forestry, and energy production have often been developed to serve national (country-wide) political interests and economic development needs. However, these national priorities often conflict with the local economy and non-industrial usage. Industrial activities in the North have damaged ecosystems and disrupted traditional livelihoods such as reindeer herding and fishing. While these undertakings bring economic benefits to some, particularly to the state, it is notable that their negative effects, such as ~~quality of~~ land degradation and the loss of access to resources, have often fallen disproportionately on Sámi communities and people (Joona & Joona, 2023a). This uneven distribution of costs and benefits is rooted in long-standing patterns of exclusion from participation in decision-making.

The second aspect of environmental and climate justice relates to effective participation in decision-making. Although the Nordic countries have strong democratic traditions and social welfare practices, decision-making on public policy and land practices has not always allowed meaningful involvement from Indigenous and local communities, resulting in various environmental conflicts (Sigeman, 2025). While official consultations with Sámi representatives take place as required by domestic legal obligations and the principle of Free Prior and Informed Consent (FPIC), in practice, the final word for major land use and development decisions is often made by national or regional authorities with limited input from those most affected (Cooper, 2023) The structures facilitating local participation sometimes also perform poorly due to how the processes are carried out (Åberg et al., 2024). This lack of community influence can undermine trust in government and reinforce existing inequalities.

The third aspect of environmental and climate justice concerns recognition, i.e., the extent to which different groups' identities, knowledge, and rights are valued in land use decision-making. Although to some extent Sámi rights are recognised along with some forms of local devolution (internal self-determination) in the Nordic countries, their policies on land use have arguably treated Sámi rights and knowledge as less important than scientific or economic considerations (Joona & Bergmann, 2024; Joona, 2023b). For example, how the Sámi people, in general, and the Sámi working within reindeer husbandry, in particular, understand and interact with the environment are frequently overlooked in established decision-making. This makes it harder for Sámi perspectives to be reflected in official policy and, consequently, more difficult for Sámi reindeer husbandry to continue. Failing to include Indigenous knowledge deepens their marginalization and prevents Nordic land use policies from fully addressing local social and environmental needs.

Nordic land-use decisions also bring to light a real dilemma: balancing global, national, and local sustainability. The green transition—the movement from fossil fuel-based energy systems—has sparked demands for energy and mineral development in the North. However, local conflicts, including Sámi rights, complicate decisions to move forward with such resource extraction. These activities contribute little to environmental justice at the local level in the Nordic North ~~local level~~ as profits from extractive industries and energy production tend to flow to the South or abroad. Yet the northern communities bear the effects of environmental degradation and long-term societal disruption that accompany such undertakings. To avoid these negative consequences, the Nordic countries must develop an improved legal framework that considers these realities and the responsibility of business and state planning actors to address them.

Amidst what Sámi representatives have termed "green colonialism," the environmental and climate justice dilemma of "who can do what with the land" also becomes a planetary concern. To consider this reality requires using another lens of justice, decolonial justice, which is only possible to employ when affected individuals, and in particular, Indigenous individuals and organisations have a direct say in decisions about land use (Chuffart, 2025). One might say that land-use rights and their concerns can operate on another international scale. What may be good for the global environment may not necessarily be good for the local environment and population (Stjernström et al., 2024).

It should be noted that promoting Sámi rights in the Arctic areas of the Nordic countries will not eliminate the global demand for minerals and 'green' energy. Instead, it might simply displace such extraction efforts elsewhere, often to the Global South, where mining operations frequently rely on similarly exploitative labour conditions, weak environmental regulations, and the violent dispossession of other local and Indigenous Peoples. This way, the burden of colonial land (ab)use may shift from one historically exploited territory to another (Stjernström et al., 2024). The challenge, then, is not only about protecting Sámi rights but about confronting the dilemma of decolonisation within a globalised extractive economy. While limiting industrial exploitation in the Nordic North may be framed as environmental progress, it can often come at the cost of creating new 'sacrifice zones' in the Global South.

These shifting geographies of harm (Zografos, 2022) found in these climate and environmental justice dilemmas also reflect some deeper structural asymmetries and issues of governance within the Nordic states. However, it is in the Arctic region that these embedded asymmetries are most apparent within contemporary governance arrangements (Wood-Donnelly, 2023b). In the context of core-periphery relations, the Nordic South rarely asks its populations to bear the costs of its resource decisions. Still, both regions remain entangled in the broader global pattern of displacing environmental and social externalities onto local spaces that have historically been relegated to the margins. This raises fundamental questions about justice and land use(s) in the Nordic North: whose lands, livelihoods, and futures are considered expendable? The justice, environmental and climate frameworks help us to frame questions regarding this displacement both in the Nordic setting and in the Global South. Within this context, there appears to be a bi-directional pull of land (ab)use and demands for justice.

10.3 Land Use, Nordic Colonialism, and Integration into Global Processes

Not only have the Nordic countries been involved in the colonisation of other places (e.g., the Caribbean or Africa), but they, themselves, have at times been occupied by invading forces (e.g. Germany or Russia) or other occupying powers (e.g. US, UK). They have also colonised their northern peripheries, including Sápmi and Greenland. Although such colonialism is often viewed as a phenomenon of the past, today, reference to a new "green colonialism" now appears in discussions in the North about the global pursuit of resources required for the green economic transition that continues to promote growth and consumption.

Beyond the existence of a core-periphery condition in overall Nordic political and economic relations, it can also be argued that there is a certain north–south logic in policy frameworks that limits land use practice to narrow strips of geography, including the Nordic North (Wood-Donnelly, 2024). This approach restricts the region's development, requiring the North to position its development relative to the South rather than to its nearest geographies in the east or west. Such positioning creates challenges in Sápmi, the traditional homeland of the Sámi People, for it is divided between four countries (Russia, Finland, Sweden and Norway) each pursuing its core-periphery needs. These can disrupt traditional economic practices and land use for reindeer herding, hunting, and fishing. In this regard, this geo-economic reality contributes to a real justice dilemma for the Sámi People and other established residents of the North.

The political and economic dynamics of the Nordic-EU relationship can also influence decisions on land use in the region. This can be seen not only in the volume of new directives coming from Brussels but in the actual trade-offs that must be made between traditional practices in the region and the need to conform to new EU policy priorities. Real tensions and conflicts can arise in attempting to reconcile established reindeer herding with the needs of new green energy production, especially concerning the introduction and siting of wind farms. The resulting distribution of harms and benefits from such efforts raises serious questions of justice embedded in core-periphery relations of various types.

Although the basic core-periphery model (Wallerstein, 1974) is useful for depicting high (core) and low (periphery) relationships and priorities, it does not fully capture what is going on in the Nordic North. The original model suggests one core, and the rest is a periphery. It also assumes that the main economic actors and resource demands all come from that core. However, the reality is more complex, as there is not one "core" but many. In one context, the EU can be seen as the core, with the Nordic states as the periphery. In another, the more densely populated areas around the Nordic capitals form a core, and the Nordic Arctic represents the periphery. However, in today's globalised world, increasing interest and pressures are emerging from other "cores" far beyond Nordic or European borders. These cores can also impact the Nordic North and the peoples who reside there. In addition, existing rivalries (e.g. Europe and China) can also influence "core" priorities, further complicating the lives of those who live in such peripheries.

A notable example of this dilemma in Scandinavia is the Násávárre (Nasafjell) mountain quartz mine. It produces a critical mineral needed for producing semiconductors, essential for many electronic devices required for the "green transformation." Although the mine was intended to reduce reliance on existing sources (i.e. from China) and lessen the burdens of development on the global south, it was to be developed in a central grazing land area of vital importance to reindeer herding in Norway and Sweden. The negative aspects of noise and dust associated with such mining would have had a major negative impact on five reindeer husbandry districts.

Despite concerns for Indigenous rights and environmental protection, the case progressed in the decision-making process until it was slowed by a controversy related to the Fosen Vindkraft case (Broderstad & Gross, 2025), where the Norwegian Supreme Court ruled in favour of Sámi rights. Another twist in the story is that the Norwegian corporation developing the mine, Elkem AS, is owned by a Chinese company and controlled by the Chinese state (Stjernström et al., 2024). This example illustrates some of the complicated relationships and difficulties with the legitimacy of land use practices and decisions made at different levels and geographies.

Another type of challenge can be examined within the context of the municipality of Kiruna in North Sweden. Residents there have long expressed concern about being used as a "resource colony" and not benefiting from such exploitation (Dagens Nyheter, 2025). The city recently used its municipal veto to deny opening another iron ore mine in the area, citing a previously never-used paragraph in the Swedish *Planning and Building Act* (Sveriges Riksdag, 2010). Arguments from the central government had emphasised the need for local economic development, more jobs, more inhabitants, etc. However, Kiruna's representatives noted they already had over 1000 unfilled jobs in the community and that "fly-in, fly-out" workers would not contribute much to the local economy's tax base (Pelling, 2024). Because Swedish income taxes are tied to home residency, if new mine employees worked in Kiruna but lived elsewhere, their income taxes would not support local costs for healthcare, infrastructure, and other programmes.

The Kiruna case is another example of the dilemma found in many Nordic planning and governance systems where land use practices become intrinsically connected to broader issues of society. Although existing market forces may make the proposed project unviable long-term, this may not end the process. This is because there are mechanisms for the state to overrule the municipality to avoid an Investor-State Dispute Settlement (ISDS) claim. A similar situation can be seen in Greenland regarding an ongoing dispute over the cancellation of licences for a uranium mine (Greenfield & Weston, 2025). Such dynamics highlight the tensions that can exist today between local land use policies and globalised networks of capital and environmental security. As local and national authorities navigate competing priorities domestically, they are simultaneously forced to reckon with external economic pressures and transnational commitments.

In Iceland, a primary concern over "land" use is, in fact, more connected to the sea and cod, one of the nation's most precious resources. There is considerable consternation over the local impacts of the Individual Transfer Quota (ITQ) system, which allocates fish catch quotas. While originally introduced in the 1990s as a mechanism to ensure the sustainability of the fish stocks in the seas around the country, over time, the system allowed for the transfer of such quotas from small fishing communities and their boats to much larger commercial fishing companies. It has been argued that the ITQ system has had detrimental consequences for Iceland's small-scale fishers and coastal communities. The system is "inextricably and irreversibly entwined with national and even global financial institutions and processes" (Chambers et al, 2020, p. 329). The result is that the wealth of Iceland is expoited not for the benefit of local citizens but for the enrichment of global corporations and their shareholders.

While land use policy should be specific to individual countries, the examples cited above show that "who can do what with the land" is affected by domestic legislation and broader economic, legal and political relationships. These cases demonstrate that many 'cores' and 'peripheries' are also divided across local, national and international scales. Questions of justice in affairs on land use and environmental decisions make the provisions of the *Aarhus Convention* important. It guarantees the right to live in a healthy environment.

The Aarhus Convention also mandates transparency in the distribution of harms and benefits in environmental decision-making. This is especially relevant for ensuring justice within the Nordic countries themselves and in their relations with other actors, such as the EU or countries with economic interests in the Nordics. While all Nordic countries are party to the convention, it is not specifically 'north', making it somewhat of a weak instrument in promoting a healthy environment in the Arctic.

10.4 Land Rights Vis-à-Vis the State

Land rights in the North, both for use and access, follow a complex history of Nordic colonisation and colonial relationships with Indigenous and non-Indigenous residents and citizens of the area. In that sense, the historical Northern expansion of these states over time, and the states' view of the North as a resource-rich place, have all played an important role in how access and use of lands in this region have been defined over time. This has led to a varied and layered structure with legal frameworks which may not align with customary practices. It has also meant that contemporary political and legal negotiations continue to influence the changing shape of land use and rights.

For instance, the rights afforded to Sámi communities and to reindeer husbandry districts, particularly those related to hunting, fishing, voting, and land use, are central to understanding these dynamics and the tension between public and private control of land and resources in this northern area. Another complicating factor is that these Sámi rights are restricted. In Sweden and Norway, they apply only to those active in reindeer husbandry, which leaves the majority of Sámi in the North without extended rights. A similar pattern occurs in Greenland, but in a wider geographic frame, given the distance to Copenhagen.

Moreover, the slow acknowledgement of Greenland's Inuit as an Indigenous People, and their right to self-determination and permanent sovereignty over their natural resources and land, has also shaped understandings and practices regarding land use there (Johnstone, 2020). Internal conflicts and broader public debates have been fuelled by ongoing tensions between conflicting claims of state sovereignty versus the customary rights of the local community and their rights to self-determination. Land use policy in Greenland will likely continue to evolve as the territory makes its way to eventual independence from Denmark.

In Sápmi,[1] national borders and citizenship rights have fragmented traditional Sámi territories and further exacerbated the alienation of Sámi communities. The imposition of state borders by Norway, Finland, and Sweden has enforced external control over Sámi mobility and cross-border uses of the land. They can restrict the Sámi's abilities to engage in traditional hunting and fishing and fully exercise their economic and cultural rights related to reindeer herding. However, one crucial document, the annex to the border treaty made between Denmark-Norway and Sweden-Finland in 1751, has played an important role in addressing these concerns. This annex, called *Lappekodicillen* (Lapp Codicil of 1751), empowers the Sámi to move across these imposed borders. This Codicil is still valid and of vital importance for the Sámi and Sápmi (Pedersen, 2006; Stjernström, 2022). It is their Magna Carta and has existed even before the eras of Nordic nationalism and state-building.

Hunting and fishing rights for the Sámi People vary considerably across Norway, Finland, and Sweden. Today, the "green transition" requirements have led to new examples of resource use and extraction in the North. Energy projects like

[1] Sápmi is the traditional homeland of the Sámi Indigenous people, which is today divided across northern Norway, Sweden, Finland and part of Russia.

wind turbines in Norway now openly conflict with established grazing land practices. Mining and hydropower development can challenge local communities and economies since dams block access to crucial pasture areas and rivers traditionally used for reindeer herding. There have also been conflicts over recent resource extraction and forestry efforts in Sweden and Finland. In Sweden, the development of new mineral deposits near Kiruna, seen as an important contribution to the green transition, has expanded into existing reindeer grazing zones. In Finland, commercial logging in the northern forests has led to disputes, as it shrinks reindeer grazing areas and disturbs traditional migration routes.

The push for expanded economic growth in all three countries has clashed with Sámi efforts to exercise their culture and traditional livelihoods. For instance, Norway recognised that Sámi hunting and reindeer herding rights are rooted in *time immemorial* usage yet remain subordinated by property-based state sovereignty arguments. Norwegian law also tends to compartmentalise Sámi rights within regulated systems that align closely with national economic and environmental objectives. This has been well illustrated by examples such as the well-known Alta case from the 1970s (concerning building a large hydroelectric dam on the Alta River) and the 2021 Fosen case concerning building wind turbines in southern Sápmi. In each case, state energy production and state-licensed use of lands have come into conflict with more traditional, small-scale economy land use practices of the Sámi.

Central to the Sámi position are their rights under Article 27 of the *International Covenant on Civil and Political Rights* (ICCPR) (United Nations, 1966). This provision protects their human rights by asserting that state actions must not undermine minority cultures without their genuine participation or consent. Further articulated through the *International Labour Organisation Convention 169* (ILO 169) and the *United Nations Declaration on the Rights of Indigenous Peoples* (UNDRIP) (United Nations, 2007), these rights provisions emphasise cultural autonomy, participatory governance, and sustainable land use practice. However, among the Nordic countries, only Norway has ratified *ILO 169* and the UNDRIP, and even though there has been some level of implementation, these international agreements are not legally binding. However, this does not mean there has been no legal progress.

For instance, in 2020, the Swedish Supreme Court recognized the Girjas Saami Reindeer Husbandry Association's right to manage hunting and fishing on certain Crown lands. This was based on the principle of immemorial prescription, emphasising that the State had not historically asserted competing claims. This particular outcome reinforces the significance of Sámi autonomy within traditional resource management and helps clarify the status of Sámi land rights under Swedish constitutional and international legal frameworks (Lundmark, 2024).

As noted, a recurring theme across the northern section of Scandinavia has been the destruction of Sámi small-scale economies, predominantly driven by extractive industries and large-scale economic developments. These industries often encroach upon Sámi lands and undermine reindeer herding practices. In addition, climate change further compounds this problem. It disrupts the traditional migratory patterns of reindeer herding and effectively erodes cultural sustainability. Consequently, the Sámi are compelled to participate in an imposed economic structure incompatible with traditional ecological knowledge and land-use practices. As a result, Sámi's right to engage in sustainable economic practices is continually compromised by state interests and the economic imperatives of multinational corporations. While international human rights frameworks theoretically protect traditional land uses, their practical enforcement remains problematic within the Nordic region.

In the Northern Norway county of Finnmark, the *Finnmark Act of 2005* emerged as an important legislative tool designed to protect Sámi rights and cultural practices. It transferred control of land previously managed by the Norwegian state to a new special organisation called the Finnmark Estate. This organisation includes representatives from the Sámi Parliament of Norway and the local county council. Under the Act, about 95% of the land in Finnmark County is managed with respect for Sámi traditional land uses, especially reindeer herding, hunting, and fishing. However, as Kuokkanen notes, the *Finnmark Act* only partially recognises Sámi land rights in Finnmark (Kuokkanen, 2019). While this law gives Sámi and local stakeholders some control over land management, it does not fully satisfy the Sámi's broader goals for self-determination.

Sámi hunting and fishing rights were first codified in Sweden by the Reindeer Herding Act of 1886. This legislation protected and limited rights to traditional Sámi livelihoods, primarily reindeer herding. Consequently, hunting and fishing rights become contingent upon having a Sámi identity as defined through historical and economic practice rather than broader conceptions of Indigenous self-determination. Such an approach has often placed Sámi interests secondary to broader national development initiatives, resulting in frequent challenges to traditional Sámi economies and lifeways.

In Finland, the Sámi possess more clearly delineated voting rights through the Sámi Parliament. It was designed to ensure Sámi autonomy within state frameworks. However, a challenge remains in the practical execution of their rights to hunting and fishing. These are continually undermined by conflicting state interests, particularly those related to extractive industries and broader economic development agendas. Given the complexity of voting rights and representation within the Sámi Parliament in Finland, pursuing Sámi land use rights has not been high on its agenda. Nor has it been in the Finnish Saami Parliament's interests to gain more autonomy (Joona & Joona, 2023b). As a result, it is unlikely that the Parliament will push for more formal recognition of these rights in the future.

Land ownership and public versus private land use are important issues in state-licensed activity. State approaches to land use vary significantly among Norway, Finland, and Sweden. Norway, emphasizing public control of land, implements comprehensive regulations regarding Sámi use rights, and often subordinates Indigenous interests to broader public or national economic strategies. Sweden similarly restricts Sámi land use rights within frameworks of state ownership and national economic interests. It emphasizes conserving interests in nature, outdoor life, husbandry, and mineral development as part of the national interest (Geological Survey of Sweden, 2020). This approach diminishes Sámi autonomy in managing their resources and territories, even when their rights to these resources are understood as equal under the law or when existing law suggests that the most sustainable use should be prioritised.

Although the Sámi parliaments were created to foster autonomy in the Nordic North, they seldom deal directly with land use issues. Norway's 1987 *Sámi Act* established its Sámi Parliament. Article 108 of the Norwegian Constitution also mandates state support for Sámi language and culture. Sweden's Sametingslag (1992) and Finland's *Act on the Sámi Parliament* (974/1995) assign cultural and administrative duties to Sámi bodies, yet these often operate only in a limited capacity. Norway, Sweden, and Finland follow a pattern of state-driven devolution that recognizes Indigenous identity through legislation but remains limited in scope by national legal and political structures. These processes define how far Sámi self-determination can go within each state and confine devolution spatially to the North without giving Sámi control over the land.

In practice, the state argues that doing so would imply that the nation-state had given up its authority over national territory. However, this runs counter to current international law on Indigenous peoples' right to *internal* self-determination. Nordic states, such as Sweden, argue that the customary rights of Sámi (sedvanerätten) are equally strong as property rights but do not necessarily require recognizing full Indigenous ownership of land. The approach allows the state to maintain sovereignty over territory by balancing customary usage against other public and private interests. This reveals ongoing tensions between Nordic state-centric models of ownership and land use and the idea of Indigenous self-determination.

10.5 Land Use Policy and Degrees of Exceptionalism

Nordic land use policies reflect some degree of internal colonialism and state-level control over regional interests. This is particularly seen in land use policy and governance provisions towards Sápmi and Greenland. This policy arena can be characterised as diverging in forms and degrees of protection and, in some instances, a lack thereof among the Nordic countries. This character is partly explained by the mixed demographic structure of Sápmi, which has shaped Indigenous rights-based discourse across Norway, Sweden, Finland and Denmark's relationship with Greenland. In the latter area, there are no special provisions for Indigenous rights in the *2009 Greenland Self-Government Act.*

In Sweden, Sámi governance arises from a legacy of centralized state control. Historical processes of land dispossession, most notably through the *Avvittringen* (land partitioning) between 1873 and 1925, formed part of a colonial strategy whereby the Swedish Crown expropriated vast areas of Sámi land in the north. Despite such history, the Swedish state continues to deny the need for full restitution while maintaining that the Sámi hold customary rights rather than outright ownership of the land.

Sweden is not a party to *ILO 169*, which would better protect Sámi rights to land. Norway, in contrast, is a party to ILO 169 and has been more proactive in implementing its legal obligations toward the Sámi. However, despite this formal recognition and other constitutional protections within Norway, challenges continue in implementing Indigenous rights, particularly when balancing state economic interests with local governance. In Finland, rights connected to reindeer husbandry are complex as this form of economy is practised by both Finns and Sámi herders.

Most Nordic land use legislation has tended to box the Sámi people into a reindeer herding category. However, such categorization raises questions concerning who should be entitled to what is often called 'special rights.' Only a small minority of the Sámi population today engages in culturally significant economic practices across Sápmi, such as small-scale traditional reindeer herding. Likewise, most Sámi now live outside of Sápmi, and non-Sámi populations within the region complicate narratives of historical justice. This all leads to complex negotiations over land use, cultural recognition, and coexistence.

In Sweden, the debate around land use policy and human rights revolves around whether the Sámi People should be recognized as equal citizens with the same rights as others or hold distinct, constitutionally protected rights. The Swedish 'rights' issue is also compounded by land restitution: Should the state enforce land reparations, or should it pursue alternative means of protecting a culture at risk of disappearing?

What is clear, however, is that the governance of land and resources across the Nordic region reflects the interplay between state public policies (see Table 1) and economic imperatives. For example, national environmental codes apply to these countries with minor exceptions. However, forestry and agriculture can fall outside this framework. This frequently leads to tensions between Sámi land-use traditions and state-sanctioned extractive industries. In this case, devolution, Sámi internal self-determination, and rights are insufficient to secure land use rights, which emerge in various public policy arrangements. In Norway, the land use policy is a hybrid between national guidelines and region-specific laws (with the Finnmark Act as the exception). Finland, like Sweden, maintains a centralized legal system but retains an older, Swedish-modelled property law (*Jordabalken*), reflecting its historical union with Sweden.

The case of Greenland provides an example of genuine devolution in land use policy through the *2009 Self-Government Act* and the *2009 Mineral Resources Act* (as amended). This legislation gives the Home-Rule Government, instead of Copenhagen, the right to control and use mineral resources on the surface and in the subsoil of Greenland. It can decide whether to grant or withhold licenses to exploit resources in Greenland. Therefore, land (ab)use and environmental protection can be discussed, contested, and settled as matters of internal politics rather than simply asserting contested Indigenous rights. It means land use decisions are ultimately made by the people of Greenland, for the people of Greenland, instead of existing as colonial and historical points of contention between Copenhagen and Nuuk.

Though Nordic countries broadly share similar land use policies concerning climate adaptation, environmental protection, cultural rights, and regional planning, they diverge in their approaches to resource management. These differences stem partly from distinct geographies, differing natural endowments, and varying political priorities tied to economic development. Nordic land use policy, however, does not fully reconcile climate and environmental goals with the effects of resource extraction. A true paradox emerges when the public policies of these countries can be seen to promote environmental governance, yet also permit activities that can undermine those same objectives.

Another unique feature of Nordic land use policy is its "Arctic Policy." It is distinctive in its dual emphasis on environmental protection and sustainable development. In this body of policy, there is a notable similarity in the tensions in the convergence between economic development, environmental security and Indigenous rights. These policies include an emphasis on climate, environment and biodiversity, sustainable economic development and business interests, and good conditions for living in the North. These policies create the underlying foundations for land use throughout the Nordic countries.

When placed together with colonial dynamics, core-periphery relations, and the drivers that connect land use and natural resources with the global economy, they can create a picture of these states' priorities. They reveal a complex interplay between state-driven economic and environmental security questions and global pressures—from climate adaptation to investment agreements. When viewed together through the lenses of rights and justice, it becomes clear that these policies struggle to balance use and abuse in land use policy.

10.6 Future Challenges

Focusing on land (ab)use in the Nordic North brings to light many unresolved tensions between climate goals, colonial histories, and local realities. What can be done to the land, and by whom, appears to differ somewhat from one Nordic state to another. This is particularly the case regarding how they manage 'their' Norths. Yet, land (ab)uses continue to be shaped by similar climate and environmental changes throughout the Nordic region.

The Nordic states share similar welfare ideals and a reputation for human rights and sustainable development. However, they greatly differ in how they address land uses relative to environmental protection, the development of 'natural resources', colonial legacies, and Indigenous autonomy. In this context, the "green transition" brings another dimension of 'green colonialism' to the Nordic North, where once again, its lands may be used in extractive ways.

In the Nordic North, wind farms and mining projects that meet global climate targets risk disrupting reindeer herding and degrading sensitive ecosystems. The failure to fully implement and protect Indigenous rights in the Nordic Arctic also raises questions about whether there is genuine recognition of their roles and voices within the development of land use policies. As the Sámi People become more dispersed, differences between reindeer herders and non-herders may also increase. Today, many Sámi live outside Sápmi and do not benefit from specialized rights that depend on reindeer herding. In Greenland, dualities between being Inuit and/ or Danish also shape the discussion of rights and identity. These gaps and overlaps in rights and identity complicate the idea of a unified Indigenous interest. It spurs internal debates over who can legitimately speak for indigenous peoples on land issues, which lands, and in what capacity.

The European Union has also emerged as an external decision-maker on environmental targets and mineral extraction in the Nordic Arctic. Nordic governments must now comply with new EU directives affecting land use in the area. They need to respond, as well, to local protests demanding fair compensation and a voice in managing resources. In theory, Nordic planning processes treat reindeer husbandry as one land-use interest among many. Yet, they are also expected to respect its deeper cultural significance and honour its practice as an established right. In practice, however, much depends on the Nordic states' willingness to offer stronger legal and economic safeguards and more transparent consultation. Failure to address local grievances and to implement Sámi rights can undermine trust in public institutions, even if global climate goals are achieved.

In that context, balancing the competing multi-scalar interests (e.g., local, national, and EU-level) requires more than the current legal framework. It also demands a broader environmental and climate justice framework. This is based on the three aspects described in the chapter. It demands a land use policy that focuses on local rights and well-being, distributes resources more equitably, and respects Indigenous internal self-determination and sovereignty. These efforts can help correct past imbalances in its formulation and implementation. Still, the issue remains of ensuring long-term coexistence among those who seek to benefit from the land, now and in the future.

In that sense, recognising the temporal dimensions of land use means engaging with the full complexity of North–South tensions within each Nordic state and the ongoing effects of EU rules, corporate interests, and the shifting preferences of global markets. Without deeper reform efforts and more inclusive land-use decisions, the future can only be expected to confront again the present reality—that of continued land abuse.

References

Åberg, H. E., Pavlova, I., Santangelo, A., Amirzada, Z., Heirman, K., & Tondelli, S. (2024). Multi-directional heritage-led knowledge exchange: learning from practice in 19 rural territories. *Heritage, 7*(3), 1689–1700. https://doi.org/10.3390/heritage7030079

Bergmann, F. (2024). An efficacious remedy for status inequality? Indigenous policies in Norway and Sweden. In *Politics, groups, and identities* (pp. 1–28). https://doi.org/10.1080/21565503.2024.2331726

Broderstad, E. G., & Gross, L. (2025). The násávárre indigenous–industry dialogue: knowledges and agency in the permit process. *Arctic Review on Law and Politics, 16*, 1–30. https://doi.org/10.23865/arctic.v16.6473

Chambers, C., Einarsson, N., & Karlsdóttir, A. (2020). Small-scale fisheries in Iceland: Local voices and global complexities. In J. J. Pascual-Fernández, C. Pita, & M. Bavinck (Eds.), *Small-scale fisheries in Europe: Status, resilience and governance* (pp. 329–349). Springer International Publishing (MARE Publication Series). https://doi.org/10.1007/978-3-030-37371-9_16

Chuffart, R. (2025). Participation trophy or agency? Indigenous rights and representation at the Arctic council. In E. Conde Perez, & C. Wood-Donnelly (Eds.), *The Routledge handbook of Arctic governance* (1st ed). Taylor & Francis Group.

Cooper, A. M. (2023). FPIC and geoengineering in the future of Scandinavia. In C. Wood-Donnelly, & J. Ohlsson (Eds.), *Arctic justice* (pp. 139–153). Bristol University Press. https://doi.org/10.51952/9781529224832.ch010

Dagens Nyheter. (2025). *Kiruna Tvingas Säga Ja till Gruva: "Betraktar Oss Som En Koloni"*. Dagens Nyheter, 27 January. https://www.dn.se/sverige/kiruna-tvingas-saga-ja-till-gruva-betraktar-oss-som-en-koloni/

Geological Survey of Sweden. (2020). *Mineral Deposits of National Interest, SGU*.

Greenfield, P., & Weston, P. (2025). Fearing toxic waste, Greenland ended. *The Guardian*, 5 March. https://www.theguardian.com/environment/2025/mar/05/greenland-mining-energy-transition-minerals-environmental-laws-uranium-rare-earth-toxic-waste-investor-state-dispute-settlement-isds-aoe. Accessed March 19, 2025.

Högsta Domstolen. (2020). *The "Girjas" case—press release*. Högsta Domstolen. https://www.domstol.se/en/supreme-court/news-archive/a-decision-on-cancellation-of-real-estate-sales-agreements/

International Labour Organisation. (1989). *C169 indigenous and tribal peoples convention no. 169*. https://normlex.ilo.org/dyn/nrmlx_en/f?p=NORMLEXPUB:55:0::NO::P55_TYPE,P55_LANG,P55_DOCUMENT,P55_NODE:REV,en,C169,/Document

Johnstone, R. L. (2020). The impact of international law on natural resource governance in Greenland. *Polar Record, 56*, e21. https://doi.org/10.1017/S0032247419000287

Joona, T., & Joona, J. (2023a). The complex relationship between forest Sámi and the Finnish state. In C. Wood-Donnelly, & J. Ohlsson (Eds.), *Arctic justice* (pp. 124–138). Bristol University Press. https://doi.org/10.51952/9781529224832.ch009

Joona, T. L., & Joona, J. (2023b). Natural resource development in the Arctic and the question of Saami land rights in Finland. In R. Johnstone, Y. Tanaka, & V. Ulfbeck (Eds.), *The Routledge handbook of polar law* (1st ed., pp. 437–455). Routledge. https://doi.org/10.4324/9781003404828-33

Kuokkanen, R. (2019). *Restructuring relations: Indigenous self-determination, governance, and gender*. Oxford University Press USA—OSO.

Lundmark, J. M. (2024). The Girjas saami case: Indigenous peoples' right to dispose freely of natural resources. In Y. Tanaka, R. L. Johnstone, & V. Ulfbeck (Eds.), *The Routledge handbook of polar law* (pp. 424–436). Routledge (Routledge handbooks in law). https://doi.org/10.4324/9781003404828

McCauley, D., et al. (2022). Which states will lead a just transition for the Arctic? A DeePeR analysis of global data on Arctic states and formal observer states. *Global Environmental Change, 73*, 102480. https://doi.org/10.1016/j.gloenvcha.2022.102480

Naess, A. (1989). *Ecology, community and lifestyle: Outline of an ecosophy* (1st ed., D. Rothenberg, Trans.). Cambridge University Press. https://doi.org/10.1017/CBO9780511525599

Norwegian Human Rights Institution. (2023). *About the wind farms on Fosen and the Supreme Court judgment, Norwegian Human Rights Institution*. https://www.nhri.no/en/2023/about-the-wind-farms-on-fosen-and-the-supreme-court-judgment/

Pedersen. (2006). *Lappekodisillen i nord 1751–1859: Fra grenseavtale og sikring av samenes rettigheter til grensesperring og samisk ulykke*. Sámi Allaskuvla. http://hdl.handle.net/11250/177104

Pelling, L. (2024) *Mining for critical materials cannot undermine trust*. https://www.socialeurope.eu/mining-for-critical-materials-cannot-undermine-trust?

Reyes, J. A. L. (2021). How different are the Nordics? Unravelling the willingness to make economic sacrifices for the environment. *Sustainability, 13*(3), 1294. https://doi.org/10.3390/su13031294

Sametingslag. (1992). *1992:1433*. https://www.riksdagen.se/sv/dokument-och-lagar/dokument/svensk-forfattningssamling/sametingslag-19921433_sfs-1992-1433/

Sigeman, G. (2025). Environmental conflicts in the Arctic and potential impacts of the EU green deal. In E. Conde Perez, & C. Wood-Donnelly (Eds.), *The Routledge handbook of Arctic governance* (1st ed). Taylor & Francis Group.

Skillington, T. (2023). Climate justice. In J. Ohlsson, & S. Przybylinski (Eds.), *Theorising justice* (pp. 155–170). Bristol University Press. https://doi.org/10.51952/9781529232233.ch010

Skorstad, B. (2023). Sacrifice zones: A conceptual framework for arctic justice studies? In C. Wood-Donnelly & J. Ohlsson (Eds.), *Arctic Justice* (pp. 96–108). Bristol University Press. https://doi.org/10.56687/9781529224832-012

Stjernström, O. (2022). *Lappekodisillen—fra frihetsbrev til konflikt, Avisa Nordland*. https://www.an.no/lappekodisillen-fra-frihetsbrev-til-konflikt/o/5-4-1563771

Stjernström, O., et al. (2024). Complexity and interdependence in local mineral extraction planning: The Nasa Mountain Quartz case in Northern Norway. *The Extractive Industries and Society, 17*, 101434. https://doi.org/10.1016/j.exis.2024.101434

Strand, R. (2024). Global sustainability frontrunners: Lessons from the Nordics. *California Management Review, 66*(3), 5–26. https://doi.org/10.1177/00081256241234709

Sveriges Riksdag. (2010). *Plan-och bygglag, 2010:900*. https://www.riksdagen.se/sv/dokument-och-lagar/dokument/svensk-forfattningssamling/plan-och-bygglag-2010900_sfs-2010-900/

Tómas, R. (2023). *Coastal fishermen disappointed, angry with minister's decision, Iceland review*. https://www.icelandreview.com/news/coastal-fishermen-disappointed-angry-with-ministers-decision/

United Nations. (1966). *International covenant on civil and political rights*. https://www.ohchr.org/en/instruments-mechanisms/instruments/international-covenant-civil-and-political-rights

United Nations. (2007). *United Nations declaration on the rights of indigenous peoples*. https://www.un.org/development/desa/indigenouspeoples/wp-content/uploads/sites/19/2018/11/UNDRIP_E_web.pdf

von Wright, G. H. (1978). *Humanismen som livshållning: Och andra essayer. Andra upplagan*. Söderström.

Wallerstein, I. M. (1974). *The modern world-system I: Capitalist agriculture and the origins of the European world-economy in the sixteenth century*. Academic Press. https://doi.org/10.1525/9780520948570

Wood-Donnelly, C. (2023a). Environmental justice. In J. Ohlsson, & S. Przybylinski (Eds.), *Theorising justice* (pp. 141–154). Bristol University Press. https://doi.org/10.51952/9781529232233.ch009

Wood-Donnelly, C. (2023b). Responsibility of and for structural (in)justice in Arctic governance. In C. Wood-Donnelly, & J. Ohlsson (Eds.), *Arctic justice* (pp. 21–35). Bristol University Press. https://doi.org/10.51952/9781529224832.ch002

Wood-Donnelly, C. (2024) Negotiating the Arctic: Sustainability, governance and environmental justice. *The Fletcher Forum of World Affairs, 48*(1). https://www.fletcherforum.org/s/Negotiating-the-Arctic-Sustainability-Governance-and-Environmental-Justice.pdf

Zografos, C. (2022) The contradictions of Green New Deals: green sacrifice and colonialism. *Soundings, 80*(80), 37–50. https://doi.org/10.3898/SOUN.80.03.2022

Corine Wood-Donnelly is Professor of International Relations and the High North in the Faculty of Social Sciences at Nord University and is the Coordinator for Nord Arctic, the university's cross-faculty platform for Arctic research. She co-leads the faculty research group GEO-Security: Governance, Geopolitics & Security and is also an Arctic Six Chair (2025–2027). She has regional expertise in the Arctic with substantive knowledge of governance frameworks, maritime law, and state practice, with her research centered on normative evolution of rules in the international system. Corine conducts interdisciplinary work from within International Relations drawing from geography, public international law and history. Her research interests include normative evolutions, particularly in Arctic sovereignty and governance, maritime search and rescue, maritime territory and resource sovereignty, performative geopolitics, values dimensions in decision-making and Arctic justice.

Romain Chuffart is the current Nansen Professor in Arctic Studies (2024–2025) at the University of Akureyri in Iceland. He holds a PhD in law from Durham University in the United Kingdom and a MA in Polar Law from the University of Akureyri. He specialises in international law and governance in the Arctic, human rights and the rights of Indigenous Peoples, international environmental law, Arctic cooperation and ocean governance. Romain is also the President and Managing Director of The Arctic Institute—Center for Circumpolar Security Studies, a 501c3 nonprofit organisation and think tank based in Washington DC, United States.

Olof Stjernström is Professor in Human Geography at Nord University and head of the department/division of History, Geography and International Relations. Previously, he was Associate Professor at Umeå University in Sweden (2000–2017) and chair of the Swedish National Committee for Geography (2015–2017). In his research Stjernström focuses on the interaction between humans and nature through studies related to land-use conflicts, spatial planning, urban development, and regional policies. Many of the cases related to the above-mentioned topics relate to the European Arctic. Stjernström is currently working on projects focusing on regional economic development and regional policies in the Nordic countries.

Keith Larson is the Director of the Arctic Centre at Umeå University. His research interests are diverse and include Arctic ecosystem ecology; impacts of environmental change on Arctic species (primarily plants, pollinators, and birds), species interactions (e.g., plants and pollinators), and ecosystems; developing methods to bring traditional ecological and Western scientific ways of knowing together to improve policy and policy outcomes; understanding the role of indigenous livelihoods in shaping ecosystems and their processes; transdisciplinary and interdisciplinary research; and contrasting theories of change to understand how to effect change.

Håkan T. Sandersen is Associate Professor at the Faculty of Social Science at Nord University, Bodø, Norway. His research interests are focused on natural resource management with an emphasis on marine issues related to aquaculture and spatial management, and increasingly also on climate change and the green transition. He teaches in the Circumpolar and Nordic Studies Programme on Knowledge of Land, Seas and Environment.

Part VI
Final Observations

Chapter 11
Summary and Conclusions

Anders Lidström, Douglas C. Nord, Gunnel Gustafsson, and Niklas Eklund

Abstract This final chapter of the book provides a summary of the previous discussions and a statement of conclusions that can be reasonably drawn from them. It points to certain cross-cutting themes that have emerged among the previous chapters. It identifies the type and extent of both internal and external influences that seem to be operative in the cases detailed. It also endeavors to outline the degree of commonality and diversity that can be found in contemporary Nordic public policies. Likewise, it seeks to draw comparisons with similar efforts on the part of other communities in the broader international setting and to suggest why the Nordic experience may be one that the rest of the global community may want to give closer examination and consideration to in the future.

Keywords Nordic · Public policy · Exceptionalism · Inner diversity · Drivers of change

This volume began with the observation that the Nordic countries are among the least observed and scrutinized societies of the modern world, despite their relevance for providing alternative public policy solutions in so many respects. The purpose of the book has been to address this lack of knowledge by providing some understanding of contemporary developments in different areas of public policy within these countries. A group of distinguished authors from several different academic disciplines across the Nordic region have offered their insights. They have considered whether the Nordic countries and their policy undertakings are "exceptional" when compared to the rest of the world. These authors have also sought to identify the extent of any "inner-diversity" that might exist among such public policy efforts within the Nordic region.

This final chapter summarizes the main findings of their research inquiries. We begin by focusing on the question of commonality and diversity. We then continue with a discussion of the important drivers of change and continuity within Nordic

A. Lidström · D. C. Nord (✉) · G. Gustafsson · N. Eklund
Umeå University, Umeå, Sweden
e-mail: douglas.nord@umu.se

D. C. Nord (ed.), *Contemporary Nordic Public Policies*, Springer Polar Sciences,
https://doi.org/10.1007/978-3-032-11292-7_11

public policy. Finally, we conclude with suggestions as to why the Nordic countries should be given further attention both within the international academic community and on the part of informed global observers.

11.1 Degrees of Commonality and Diversity

It is a debated matter among scholars whether there has ever existed a “Nordic model” of public policy, in any of the two meanings of the concept that are sketched out by Eklund and Gustafsson in Chap. 1. On the one hand, the concept arises from the observation that many features of the Nordic countries, including their public policies, appear similar to one another in a way that sets them apart from other countries of the world. As such, they may seem to be “exceptional” in character. On the other hand, the concept of the Nordic model has also been derived from a view that suggests that the Nordic societies serve as a “role model” or as an ideal inspiration for other communities to follow to in their own development of their public policies. In this regard, the Nordics are viewed also as “exceptional,” this time in a heuristic fashion.

The heyday of both interpretations came during the first few decades following the Second World War. Lidström in Chap. 2 of the book refers to this as *the tax financed expansion period* of social welfare and public services within the Nordic countries. This was the time when most of the foundations of the Social Democratic welfare state were established. It provided universal and generous benefits and public services, while still retaining a basically market driven economy. This provided “a middle way”—an alternative path to both capitalism and communism offered by two superpowers of the era, the Soviet Union and the United States.

A significant number of commonalities still characterize the Nordic societies today. A very important asset has been the extensive amount of trust that Nordic citizens continue to place in both their fellow community members and in public authorities. Combined with low levels of corruption, this attitude has allowed public decision-makers to tax their citizens at significant levels to fund major commitments in the welfare system. In addition, as underlined by Puranen and Brunkert in Chap. 3, citizens in the Nordic countries remain “distinctive” in the sense that they are highly secular and emphasize emancipative values of self-expression, individual autonomy and broad participation in decision making. These fundamental values remain as important underpinnings of Nordic exceptionalism. To the surprise of some, the citizens of these same countries generally think that they should obey their government officials, not out of blind support, but because they have been selected in a legitimate manner. Participatory democracy remains a cherished belief. Another current expression of the relevance of these regional values is that the populist parties that have emerged in the Nordic region have tended to be less anti-EU, less misogynist and less authoritarian than corresponding groups in Europe and elsewhere, although they still hold nationalist and anti-immigrant beliefs.

The Nordic welfare system was gradually reduced and became less generous and redistributive during *the liberalization and globalization phase* of policy development as described by Lidström in Chap. 2 of this book. In tandem with this, Byrkjeflot and Strang suggest in Chap. 4, that the Nordic Social Democratic parties lost much of their political hegemony within the region during this period when they changed their justification for the Nordic model from that of realizing a social democratic ideal to one that became linked to Nordic traditions associated with community life, or to a specific type of Nordic individualism. Despite its weakening, new common features of the Nordic model emerged in such areas as family and gender policies. As shown by Lindmark in Chap. 5 of this volume, a booming stock market in the Nordic countries during this second period of policy development also made it cheaper for private companies rather than the state to raise capital for public services. This contributed to a further transformation and marketization of the Nordic welfare model.

The liberalization and globalization phase of the Nordic countries also witnessed change in the public sectors of the Nordic countries through the adoption of New Public Management ideals. This belief in the wisdom of the marketplace and the effectiveness of the private firm, provided new guiding principles for how public authorities should be governed. However, they also tended to undermine lines of trust within these public authorities as they assumed that subordinate and implementing actors needed to be controlled and steered. As Bentzen and Winsford have shown in Chap. 6 of the book, this generated a counter reaction in the form of the trust reforms of Denmark, Norway and Sweden that emerged at the end of the liberalization/globalization phase. A common aim of these policy reforms was to restore the professional autonomy of public employees and to reduce excessive controls and documentation.

Although certain distinctive Nordic traits were retained throughout the entire liberalization and globalization era, any effort to maintain an alternative model for policy development proved challenging. The liberalization of trade, along with a more globally-oriented investment approach, brought the export-dependent Nordic economies into closer contact with the rest of the world and made it increasingly difficult for the Nordic countries to chart their own course. Finland and Sweden's decision to join the European Union in the early 1990s, along with Norway and Iceland being connected through various associational agreements with the EU, led to a significant integration of the Nordic region into the overall European political economy.

The Nordic countries also became more like other European countries through demographic changes that also occurred during this period. Where once the populations of the region had tended to be relatively homogenous in terms of ethnicity, religion, and languages (Finland being the one anomaly) this all changed during the final decades of the twentieth century. Puranen and Brunkert describe in Chap. 3 of the book how the growing influx of both refugees and labour immigrants from other parts of the world was to have the most profound consequences for the societal composition of the Nordic countries. In a way, their arrival made the Nordic countries less "exceptional" and more like other migrant receiving societies in Europe—including witnessing the emergence of new populist political parties in the region which were opposed to these arrivals. Another societal feature that became increasingly significant during the latter portion of the twentieth century was the growing activism of the region's Indigenous communities including the Sámi in Scandinavia and the Inuit in Greenland as described by Wood-Donnelly et al. in Chap. 10 of the book. In attempting to craft responsive policies, the Nordic states found themselves in the position of other "colonizing" societies around the globe.

The reduction of Nordic differences vis-à-vis the rest of the world have also been seen in some other key policy areas. A prime example is how Nordic security policies have changed under what we have labelled *the post-global era* (Lidström in Chap. 2). These became animated by climate change, trade conflicts and a weaker rules-based international order. In particular, the Russian full-scale attack on Ukraine in 2022 led to Finland and Sweden reconsidering their previous policies of neutrality and becoming members of NATO. As noted by Eklund and Bartoszewicz in their review of Nordic understandings of security in Chap. 7, the common threat from an aggressive Russia changed the established Nordic set of defence policies from those which featured a diversity of approach into ones with greater solidarity and convergence. They also point out that a key feature of Nordic defense policy remains one of "societal security" meaning that the whole population and the totality of national resources are involved in its development.

This fluctuating balance between commonality and diversity during this post-global era can also be seen in other policy areas. This is illustrated in Chap. 8 with regard to how the Nordic countries relate to the Arctic. There Nord and Schmidt suggest that these countries carry out their foreign policies toward this increasingly important area of the world in some similar ways. They focus on institutional building and the establishment of trust between circumpolar neighbors. Yet, at the same time, each of the Nordic states has its own unique geopolitical location, set of priorities, and established affiliations that also direct its Arctic policies. Here we can see forces of commonality and distinction operating in tandem with one another which may be an increasing evident pattern across the Nordic region with regard to all contemporary public policy. This joint strategy also seems to characterizes how the Nordics address the environmental and health challenges generated by climate change in the Arctic. Their efforts are analyzed by Evengård and Emelyanova in Chap. 9 of the book.

The entire Nordic region, and especially its Arctic areas, are also rich in raw materials that are in demand during the "green transition" of the first portion of this century. The Nordic North has an abundance of hydro and wind power along with substantial mineral deposits—including rare earths—that are deemed essential for this undertaking. As a consequence, there are strong national and global interests aimed at their development and utilization. However, the lands in which they are often found can also the homes of Indigenous peoples. As underlined by Wood-Donnelly et al in Chap. 10, while Norway, Sweden and Finland pursue similar land use policies in the North, these also have clashed with the Sámi peoples' right to self-determination. The Inuit people in Greenland, on the other hand, seem to be in a somewhat stronger position which includes the right to make more of their own decisions about land use.

In today's world it may be increasingly misleading to assume that the Nordic countries can always be viewed as a common entity. Despite often being considered together as pursuing a distinctive regional "model," significant differences continue to exist among these several national communities. Finland is not only exceptional due to its very different language but also because of its proximity over time to an authoritarian neighbor, Russia. Puranen and Brunkert show in Chap. 3 that citizens in Finland have less liberal and post-material values than are found in the other Nordic countries. Consequently, their populist party (the True Finns) have more in common with those which have developed in Eastern and Central Europe.

Sweden is an extreme case in the other direction, with its citizenry having the strongest values on individuality and self-expression in the Nordic region. At the same time, Byrkjeflot and Strang in Chap. 4, point out this is the Nordic country with the broadest social welfare system. Perhaps this is a consequence of Sweden also having the longest period of Social Democratic party dominance within the Nordic region, from the 1930s to the middle of the 1970s.

Norway stands apart from its Scandinavian neighbors not only because of its geographic position along the Atlantic rim, but also due to its extensive oil resources which has given the country a unique economic status envied by the other Nordic communities. As noted by Lindmark in Chap. 5, and Bentzen and Windvold in Chap. 6, this unique feature of Norway has allowed the country a degree of latitude to insulate its public policies from broader global forces not found in the other the other Nordic countries.

Denmark combines a more continental orientation with a more liberal tradition that often sets it apart from the other Nordics in its policy development. This is pointed out by Nord and Schmidt in Chap. 8 of the book with respect to the development of its overall approach to Arctic foreign policy. Its current control over Greenland and the Faroe Islands as autonomous parts of the country, also makes it a distinctive "colonial" nation among the broader Nordic group.

Finally, Iceland in many ways retains its status as a true Nordic outlier, both literally and figuratively. It has a geographical position far from the other Nordic countries and a different type of economy which has been based almost exclusively on fishing. Its orientation as observed by Nord and Schmidt in Chap. 8 tends to be inwardly focused. Puranen and Brunkert in Chap. 3 point also to its less prominent role in broader Nordic social movements.

Nevertheless, while differences between the Nordic countries persist, there is not very much evidence presented in the chapters of this book that suggest that that these will grow over time. Many of these continuing distinctions reflect long-standing geographic factors, historical inheritances, and specific resource endowments. More recent developments, both during the liberalization/globalization phase and the contemporary post-global period, suggest that the Nordic countries are becoming more similar to one another in their public policy choices, more united in their societal attitudes and orientations, and, in many ways, increasingly like their western European neighbors. Our overall conclusion is that we are moving towards a direction of less diversity among the Nordic countries in their public policy formulation, but not in a manner that emphasizes their "exceptionalist" character. The Nordic

Model—especially as it was presented in the 1950s and 1960s—may not be of great assistance in understanding these most recent trends.

11.2 Drivers of Change and Continuity within Nordic Public Policy

It is clear that the direction and content of Nordic policymaking has been affected by both external and internal forces. Many of the more recent converging tendencies among the Nordic countries, and between the Nordics and the rest of the world, are driven by external circumstances. The Nordic societies today are all highly dependent on international trade and the general economic situation in the rest of the world. As Lindmark notes in Chap. 5, they became more exposed to these influences during the liberalization and globalization phase of their development which reduced the scope for independent Nordic solutions. Additional external pressures emerged during the post-global period when climate change and the Covid pandemic pointed to the need for joint responses and worldwide actions. Likewise, as pointed to by Eklund and Bartoszewicz in Chap. 7 of the book, the Russian aggression towards Ukraine and the increased dismantling of the established world order by the Trump administration has added to this pressure from outside and required that new types of public policy be developed in the Nordic area.

Other policy changes have come from a different source. During the post-global period, the demand for changed public policies has arisen also from internal sources including those which call for a reversal of government actions taken during the liberalism/globalization period. For example, as pointed out by Puranen and Brunkert in Chap. 3, today's support for populist parties in the region may be seen, in part, as a reaction from rural and working-class voters who have felt abandoned by the Nordic welfare state and who believe that existing public policies have benefited only the newcomer or the well-educated middle class in the big cities. Likewise, as observed by Bentsen and Windvold in Chap. 6, the current need to restore trust within the public sectors of the Nordic states can be clearly linked to the New Public Management policies from previous decades which encouraged a separation between managers and subordinate workers.

Regardless of the source change, there are also certain strong historical and institutional legacies that contribute to the distinctive character of the Nordic societies and the types of public policy that they favor. This degree of continuity within changing circumstances is remarked upon by Byrkjeflot and Strang in Chap. 4 of the book. Despite the impact of new external and internal forces that would suggest a decline, high levels of societal trust remain among the Nordics and continue to color their policy preferences. This point is also emphasized by Bentzen and Windvold in Chap. 6. The same can also be said regarding the Nordics' continuing strong commitment to emancipative and secular values that still shape their public policy choices. Puranen and Brunkert in Chap. 3 of the book stress this point. The Nordics have also continued to stand close together because of the extensive regional institutional intertwining that continues today as noted by Eklund and Gustafsson in Chap. 1. These inter-Nordic institutions and agreements remain relevant despite changes in the broader international world order. They represent a common foundation of values and objectives that continues to unite the Nordic countries.

On the other side of the coin, it is also true that the Nordics have a common history of internal colonialism in relation to Indigenous people in the Sápmi and in Greenland. This inheritance has led to multiple conflicts over land use as pointed out by Wood-Donnelly et al in Chap. 10 of the book. These are likely to continue as pressure for access to needed mineral and energy resources during the "green transition" come face to face with demands by the Nordic Indigenous communities that their local economies and ways of life be respected.

11.3 How the Nordic Experience is Relevant to the Rest of the World

The Nordic experience has been one of flexible adjustment to changing circumstances derived from a position of common underlying values and goals. Some of their public policies have remained relatively distinct whereas others have converged and become more like those carried out in other modern societies. Although certain underlying values of high trust, self-expression and secularization continue to set them apart, many informed observers can still look to the Nordic region for useful ideas when considering how contemporary societal problems can be addressed.

Such Nordic innovations include family and gender policies that make it easier for women in the region to combine the raising children with a modern working life. Another is the Nordic strategy for building trust within public authorities that seeks to enhance creativity and problem-solving capabilities among its employees. Similarly, the Nordic approach to foreign policy, one of establishing stable international institutions and resolving disputes within them, may be even more relevant today given the current global situation. Also, the Nordic emphasis on building societal security as a way to respond to contemporary economic, environmental and defense needs may well be of interest to many other parts of the world. The important point to be made here is that while the once fabled Nordic Model is no longer a touchstone for public policy development everywhere in the world, the policy efforts of the region really should be better understood and studied by both those who are within the Nordic area and those outside.

The initial observation of this volume, that there continues to be a lack of knowledge about the Nordic countries and their behavior, also calls for additional research with the purpose of exploring why some Nordic policy areas have retained a degree of exceptionalism while others have generally converged toward a more western European or global standard. Likewise, additional work needs to be done to tease out the reasons for continuing degrees of inner diversity in public policymaking among the Nordics. This will require more comparative studies utilizing both historical and contemporary perspectives and taking advantage of both qualitative and quantitative methods of analysis. Researchers from across the Nordic region, and from outside, should be involved in this effort. Also, as has been shown in this volume, research on these matters should be multi-disciplinary in character since no single discipline holds all of the answers to the questions of concern. The hope shared by all the authors of this volume is that the Nordic region will be both better understood in the future and that its public policy choices more regularly compared with other communities across the globe.

Anders Lidström is Professor Emeritus of Political Science at Umeå University, Sweden. He specializes in local, regional and urban democracy and governance. Anders Lidström was the convenor of the ECPR standing group on local government and politics from 2010–2016; Head of the Department of Political Science, Umeå University from 2010–2016 and Chair of the Swedish Political Science Association from 2015– 2017. He is currently the Swedish member of the Group of Independent Experts on the European Charter of Local Self-Government, Council of Europe. Recent publications include *Multilevel Democracy: How Local Institutions and Civil Society Shape the Modern State* (with Jefferey Sellers and Yooil Bae), Cambridge University Press, 2020; a special issue of *Journal of Urban Affairs* on "The Citizens in City-Regions" (No 1, 2018, edited with Linze Schaap) and, Sverigedemokraternas regionala politiska kultur i Skåne och Blekinge. Vad ligger bakom partiets exceptionella regionala framgångar? (Umeå Working Paper in Political Science, 2025:1--with a summary in English).

Gunnel Gustafsson is Professor Emerita in Political Science at Umeå University. She served as Director of NordForsk from 2010–2017. From 2005–2010 she was Deputy Director General of the Swedish Research Council and prior to then, Pro Vice Chancellor of Umeå University. She has also served on several editorial boards, peer research review committees and contributed to the work of a number of research financing agencies including the Bank of Sweden's Tercentenary Foundation. She has written widely on topics related to Swedish and Nordic public policy with a special focus on education, social welfare and democracy. She has also led research efforts aimed at studying local government reform in Sweden, women's political leadership in that country, and Nordic efforts to secure gender equality.

Niklas Eklund is Professor of Politics and Public Administration at the Department of Political Science, Umeå University, and the current Director of the European CBRNE Center. He also serves on the steering committee for the Arctic Centre at Umeå University and on the board of the local twenty-first Century Scandinavian Policy Studies Foundation. In his research, he specializes among other things in Arctic security, geopolitics and Russian studies. He was appointed Arctic Six Chair in Nordic and Arctic Security for 2024–2026, a position from which he is actively trying to promote transboundary Nordic research initiatives driven by geopolitical change and emergent security needs in the European High North.

MIX
Papier aus verantwortungsvollen Quellen
Paper from responsible sources
FSC® C105338

If you have any concerns about our products,
you can contact us on
ProductSafety@springernature.com

In case Publisher is established outside the EU,
the EU authorized representative is:
Springer Nature Customer Service Center GmbH
Europaplatz 3, 69115 Heidelberg, Germany

Printed by Libri Plureos GmbH
in Hamburg, Germany